COL. BELOIT IN REP. WISCONSIN U.S.
SCIENTIA VERA CUM FIDE PURA

Geological Society of America
Memoir 181

Ruptures of Major Earthquakes and Active Deformation in Mongolia and Its Surroundings

I. Baljinnyam
A. Bayasgalan
B. A. Borisov
Armando Cisternas
M. G. Dem'yanovich
L. Ganbaatar
V. M. Kochetkov
R. A. Kurushin
Peter Molnar
Hervé Philip
Yu. Ya. Vashchilov

1993

Published by The Geological Society of America, Inc.
3300 Penrose Place, P.O. Box 9140, Boulder, Colorado 80301

Printed in U.S.A.

GSA Books Science Editor Richard A. Hoppin

Library of Congress Cataloging-in-Publication Data
Ruptures of major earthquakes and active deformation in Mongolia and its surroundings / I. Baljinnyam . . . [et al.].
p. cm. — (Memoir ; 181)
Includes bibliographical references and index.
ISBN 0-8137-1181-9
1. Earthquakes—Mongolia. 2. Geology, Structural—Mongolia.
I. Baljinnyam, I. II. Series: Memoir (Geological Society of America) ; 181.
QE537.2.M65R87 1994
551.8'7'09517—dc20 93-17406
CIP

Authors' addresses:

I. Baljinnyam and L. Ganbaatar
Seismograph Station, Institute of Geology, Academy of Sciences of Mongolia, Ulaanbaatar 210351, Mongolian People's Republic
A. Bayasgalan
Laboratory of Remote Sensing, Institute of Physics and Technics, Academy of Sciences of Mongolia, Ulaanbaatar 210351, Mongolian People's Republic
B. A. Borisov
Institute of Physics of the Earth, Russian Academy of Sciences, B. Gruzinskaya, 10, Moscow, Russia
A. Cisternas
Institut de Physique du Globe, Université Louis Pasteur, Strasbourg, France
M. G. Dem'yanovich, V. M. Kochetkov, and R. A. Kurushin
Institute of the Earth's Crust, Siberian Branch, Russian Academy of Sciences, Lermontov Street, 128, Irkutsk 664033, Russia
P. Molnar
Department of Earth, Atmospheric, and Planetary Sciences, Massachusetts Institute of Technology, Cambridge, Massachusetts 02139
H. Philip
Laboratoire de Géologie Structurale, Université Scientifique et Technique du Languedoc, Montpellier, France
Yu. Ya. Vashchilov
Regional Geophysics Laboratory, Northeastern Interdisciplinary Research Institute, Far Eastern Science Center, Russian Academy of Sciences, Magadan 685000, Russia

10 9 8 7 6 5 4 3 2 1

Contents

Geological Society of America
Memoir 181
1993

Ruptures of Major Earthquakes and Active Deformation in Mongolia and Its Surroundings

ABSTRACT

In this century, western Mongolia and the area adjacent to it in China have been one of the most seismic intracontinental regions of the world. Four earthquakes with magnitudes (M) ⩾ 8 have occurred. Average displacements of several meters along ruptures more than 100 km long characterize all of them. The dominant style of faulting for each was strike-slip: left-lateral on easterly trending planes in the 1905 Bulnay and Tsetserleg earthquakes and in the 1957 Gobi Altay earthquake and right-lateral on a north-northwesterly trending Fu-yun fault in 1931. The ruptures associated with these earthquakes, with most other, smaller earthquakes, and with one older great earthquake suggest that western Mongolia is undergoing conjugate strike-slip deformation. Equivalently, the region undergoes northeast-southwest shortening and northwest-southeast extension. The component of shortening can be seen as a manifestation of the convergence between India and Siberia. The component of extension seems to mark a transition from an area of largely crustal shortening in China to another, the Baikal Rift system, where crustal extension is dominant. The average rate of seismic deformation in western Mongolia in this century consists of 49 (± 15) mm/a of northeast-southwest shortening and 40 (± 12) mm/a of northwest-southeast extension. Such high rates imply strongly that the twentieth-century seismicity has been abnormally high. Moreover, crude estimates of average recurrence intervals for great earthquakes on the Bulnay fault and in the Gobi Altay region are about 1,000 yr. Approximate Holocene or late Quaternary average slip rates on these faults are a few millimeters per year, suggesting that Mongolia is being sheared left-laterally with respect to Siberia at about 10 mm/a (between 5 and 20 mm/a). A similarly crude estimate for right-lateral slip along the northwest-trending Mongolian Altay is also 10 mm/a. We suspect that this right-lateral shear is a manifestation of the left-lateral regional shear parallel to east-west planes and of counterclockwise rotation of both the Mongolian Altay and the strike-slip faults within the range. Correspondingly, the eastward translation of western Mongolia with respect to Siberia manifests itself as crustal extension and rifting along the Hövsgöl and Baikal rift zones. In the Hangay, the broad upland in the interior of western Mongolia, scattered minor normal faulting with no obvious preferred orientation appears to be the common style of deformation. This area seems to be underlain by the same relatively hot upper mantle that underlies the Baikal rift system. Thus, as others have suggested, the collision between India and Eurasia does not appear to be the only cause of the active tectonics of western Mongolia. The perturbations to the stress field in the crust resulting from the emplacement (and upwelling) of hot material beneath the Baikal area and the Hangay and Hövsgöl uplands also play a key role in the active tectonics. Finally, the active deformation in western Mongolia appears to be young. Some surface faulting bears no obvious relation to the present topography. The very flat summit of Ih Bogd, the highest peak in the Gobi Altay, may have risen from the surrounding lowlands since only 1 Ma. Thus, the rapid deformation in western Mongolia seems to have begun tens of millions of years after India collided with Eurasia, perhaps as recently as a few million years ago.

Baljinnyam, I., Bayasgalan, A., Borisov, B. A., Cisternas, A., Dem'yanovich, M. G., Ganbaatar, L., Kochetkov, V. M., Kurushin, R. A., Molnar, P., Philip, H., and Vashchilov, Yu. Ya., 1993, Ruptures of Major Earthquakes and Active Deformation in Mongolia and Its Surroundings: Boulder, Colorado, Geological Society of America Memoir 181.

INTRODUCTION

Western Mongolia and its surroundings are undergoing rapid deformation. At least in part, this occurs in response to the continued penetration of India, roughly 2,000 to 3,000 km south-southwest of Mongolia (Fig. 1), into the rest of the Eurasian continent (e.g., Molnar and Tapponnier, 1975). Largely strike-slip faulting and also both reverse (or thrust) and normal faulting play key roles in the active tectonics of Mongolia. Four earthquakes with magnitudes ⩾8 have occurred within or just west of Mongolia in this century, and their surface ruptures comprise some of the most impressive examples of recent faulting on earth. These surface ruptures and the traces of smaller earthquakes in this century and of preinstrumental events provide the most definitive evidence for the active deformation of the region.

One of our purposes here is to provide a summary of the surface ruptures of the major earthquakes. Much of this summary is drawn from the detailed work of Florensov and Solonenko (1963), Khil'ko and others (1985), and Natsag-Yüm and others (1971), but we augment their descriptions with observations that we made during three- and four-week reconnaissance trips in 1990 and 1991 and with a review of seismological studies of many of these earthquakes.

This summary is then used to infer preliminary estimates of rates of Holocene deformation and rates of recurrence of great earthquakes in parts of western Mongolia. To compare seismological measures of the sizes of these earthquakes with inferences of amounts of slip and lengths of rupture zones made in the field, we use the seismic moment (Aki, 1966):

$$M_o = \mu A_o \Delta u$$

where μ is the shear modulus, assumed to be 3.3×10^{10} N/m, A_o is the area of the fault that ruptured, and Δu is the average displacement. To compare and combine ruptures, we assume that all extended to a depth of 20 km.

As many readers may use this paper more as an encyclopedia of active surface faulting in Mongolia than as an interpretation of observation, we alert them to some easily overlooked

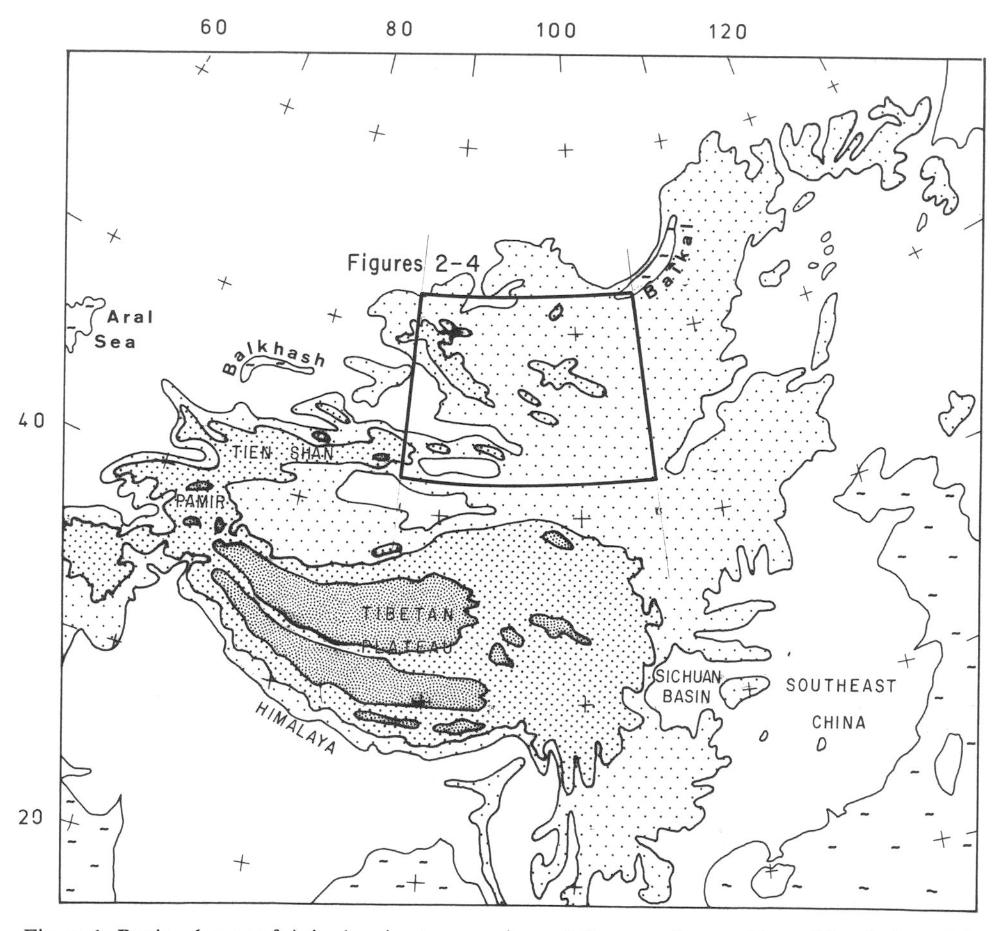

Figure 1. Regional map of Asia showing topography and the area discussed here (Figs. 2, 3, and 4). Shading, in decreasing order of darkness, shows areas higher than 5,000 m, areas 5,000 m to 3,000 m, and areas 3,000 m through 1,000 m. White areas are lower than 1,000 m. Dashed regions show large bodies of water (oceans and large lakes).

aspects. We rely heavily on observations of surface ruptures of large earthquakes to draw inferences about rates and styles of deformation. It is likely, however, that ruptures from some types of faulting leave clearer surface breaks than others. For instance, thrust and reverse faults do not always break the surface, as was the case for the Coalinga, California, earthquake in 1983 (Stein and King, 1984). Where reverse-fault surface ruptures do reach the surface, deformation can be deceptively complex, a fact well illustrated by the surface ruptures of earthquakes in the El-Asnam region of Algeria in 1981 and at earlier times (e.g., Avouac and others, 1993; King and Vita-Finzi, 1981; Philip and Meghraoui, 1983) and by the Spitak earthquake of 1988 in Armenia (Philip and others, 1992). Such complexities may be responsible for the paucity of observations of reverse faulting in the Mongolian Altay of western Mongolia, where geologic evidence attests to such deformation in Cenozoic time. Similarly, surface ruptures may be better preserved in some areas than in others. In particular, we suspect that long durations of frozen ground, and in some cases permafrost, preserve surface deformation that might be destroyed more rapidly in warmer, moister climates. Finally, of course, the seismicity of the twentieth century need not be, and for Mongolia probably is not, representative of a longer interval of time.

Because the active tectonics is affected by the earlier geologic history of the region and by lateral variations in topography, crustal thickness, and upper mantle structure, we review briefly what is known about these before examining the various surface ruptures.

To discuss the faulting in and around Mongolia (Fig. 1), we subdivide the area into separate regions and consider them individually, beginning with the Mongolian Altay and the adjoining part of western China. Then moving counterclockwise around the country, we discuss the Gobi Altay, the Hangayn Nuruu (or Hangay), and northern Mongolia (Fig. 2).

Finally, because so little has been published in English about Mongolia, we have appended a short glossary of Mongolian geographical terms, most of which we have not translated in the text, and a brief guide to the pronunciation of Mongolian words transliterated into the Latin alphabet.

REGIONAL GEOLOGY, TOPOGRAPHY, AND DEEP STRUCTURE OF WESTERN MONGOLIA

Brief summary of the pre-Cenozoic geologic history of Mongolia

The pre-Cenozoic geologic history of Mongolia is dominated by three major orogenic phases: late Precambrian in northern Mongolia, early Paleozoic across central Mongolia, and late Paleozoic through southern Mongolia (e.g., Burrett, 1974; Li and others, 1982; Sengör and others, 1988; Zaitsev and others, 1974; Zhang and others, 1984). The late Precambrian Baikalian belt lies parallel to Lake Baikal north of Mongolia (Fig. 1) and trends east-west across northernmost Mongolia. In contrast to the folding that characterizes the deformation of rock in this belt, the active tectonics of northernmost Mongolia is dominated by three parallel, northerly trending grabens, which collectively form the Hövsgöl graben system, the southwest continuation of the Baikal rift system (Figs. 2, 3, and 4). The early Paleozoic belt wraps around the Baikalides, trending northwest obliquely across the Mongolian Altay in northwestern Mongolia, roughly east-west across the southern part of the Hangay in west-central Mongolia, and northeast in eastern Mongolia. The late Paleozoic belt lies roughly parallel to and south of the early Paleozoic belt. Each of these is probably associated with subduction of oceanic lithosphere and the collision of fragments of thick (apparently continental) crust to the ancient southern margin of the Eurasian continent (e.g., Burrett, 1974; Li and others, 1982; Sengör and others, 1988; Zhang and others, 1984).

The basement of western Mongolia seems to have been consolidated by the end of the Paleozoic era. Although some folding occurred during the Mesozoic era (e.g., Devyatkin and Shuvalov, 1990; Florensov and Solonenko, 1963; Kozhevnikov and others, 1970), Mesozoic tectonic activity was notably less intense than earlier or than at present. Note that this statement does not apply to eastern Mongolia, where marine sediment was deposited and subsequently folded and where Mesozoic plutonism was widespread (e.g., Nagibina, 1975). Sengör and others (1988) suggested that much of easternmost Asia east of western Mongolia was sutured to the rest of Mongolia in the Mesozoic era.

Although this geologic history surely left Mongolia with zones of weakness and an anisotropic crust, the present deformation and its manifestations in topography seem to bear less relationship to these pre-Tertiary structures than they do to the present upper mantle structure. We suspect that modifications to the thermal structure of the upper mantle may have affected the distribution of active deformation as much as ancient structural weaknesses have.

Present topography and its relation to Cenozoic deformation

Virtually all of western Mongolia lies above 1,000 m, and the elevation of much of the region exceeds 2,000 m (Fig. 2). Because work must be done against gravity to create high elevations (e.g., Artyushkov, 1973; England and Houseman, 1989; Molnar and Lyon-Caen, 1988), variations in regional elevations profoundly affect the locations and styles of active deformation.

Trending northwest along the western margin of Mongolia, the Mongolian Altay forms the country's largest and highest range: 200 to 300 km in width, roughly 500 km long, an average of 2,260 m high (Khil'ko and Kurushin, 1982, p. 41), and with many summits above 4,000 m. The range is laced with northwesterly trending right-lateral strike-slip faults (Fig. 3). These faults bound blocks with remarkably flat-topped mountains at heights of 3,000 to 3,300 m, which Khil'ko and Kurushin (1982, p. 47) described as "semi-horsts," and basins with asymmetric cross sections. The flat tops of the mountains are commonly

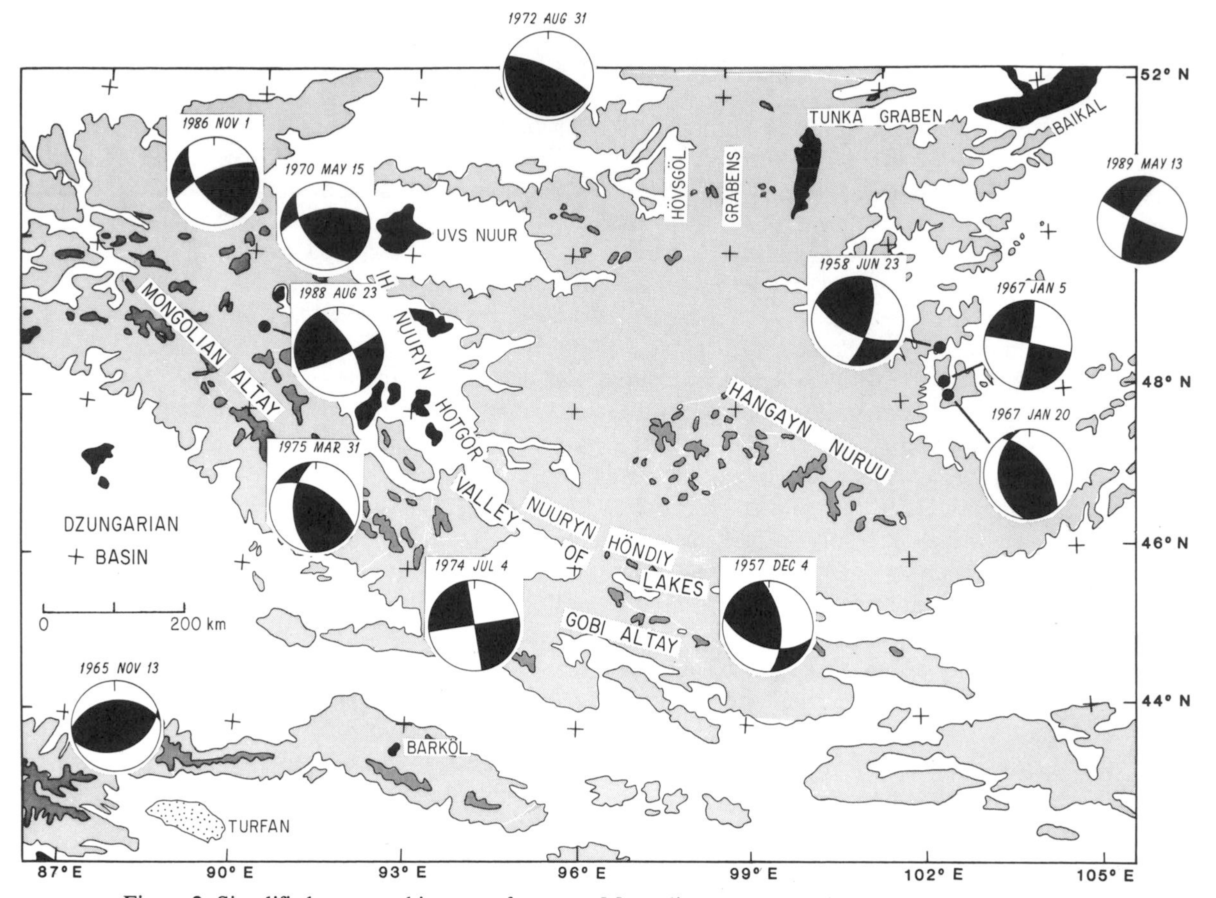

Figure 2. Simplified topographic map of western Mongolia and surrounding regions showing major geographic features and fault plane solutions of earthquakes large enough to be studied using the World-Wide Standardized Seismograph Network or its predecessors. Darkest shading indicates lakes. Gray regions lie above 3,000 m, and lighter shading shows areas higher than 1,500 m. Dotted area in the lower left is the Turfan Depression, below sea level. In lower hemisphere projections of focal spheres, the darkened quadrants contain compressional P-wave first motions and orientations of extensional strain; white quadrants include dilatational first motions and orientations of compressional strain.

ascribed to a late Cretaceous or Paleogene planation surface and subsequent Neogene and Quaternary faulting and vertical components of displacement. The basins are also high, at 1,500 to 2,200 m, and are underlain by only thin sediment (<200 m) (Khil'ko and Kurushin, 1982, p. 48). Thus, relatively small vertical components on these predominantly strike-slip faults have played a key role in shaping the topography. The high mean elevation, however, must be due to some other process. The range seems to have been built by crustal shortening associated with reverse and thrust faulting.

Mountainous terrain continues east-southeast from the Mongolian Altay as a series of elevated terrains collectively constituting the Gobi Altay. Defining the geographic boundary between the Mongolian and Gobi Altay is somewhat arbitrary. We follow Timofeev and Nikolaeva's (1982, p. 65) definition, in which the southern end of the Mongolian Altay is taken to be just south of the Shargyn Tsagaan basin, where mountain ranges trend east-west (see Figs. 23, 24, and 25 below). This is a convenient boundary because, unlike the situation in the Mongolian Altay, left-lateral strike-slip faulting, with components of reverse faulting, occurs on easterly trending planes that bound the mountains of the Gobi Altay. The opposite senses of strike-slip displacement make it difficult to treat the Mongolian Altay and the Gobi Altay as tectonic continuations of one another.

Although geomorphologists commonly consider the landscapes of the Mongolian Altay and the Gobi Altay to be different from one another, in some respects their similarities are more notable. The Gobi Altay also includes mountainous blocks with remarkably flat tops above 3,000 m, assigned to late Cretaceous–Paleogene peneplanation, and basins that stand high (Timofeev and Nikolaeva, 1982, p. 79–81). As in the Mongolian Altay, these basins and mountains are separated by faults on which large strike-slip components occur. Tectonic activity in the Gobi Altay is thought to have begun in late Pliocene-Quaternary time (Timo-

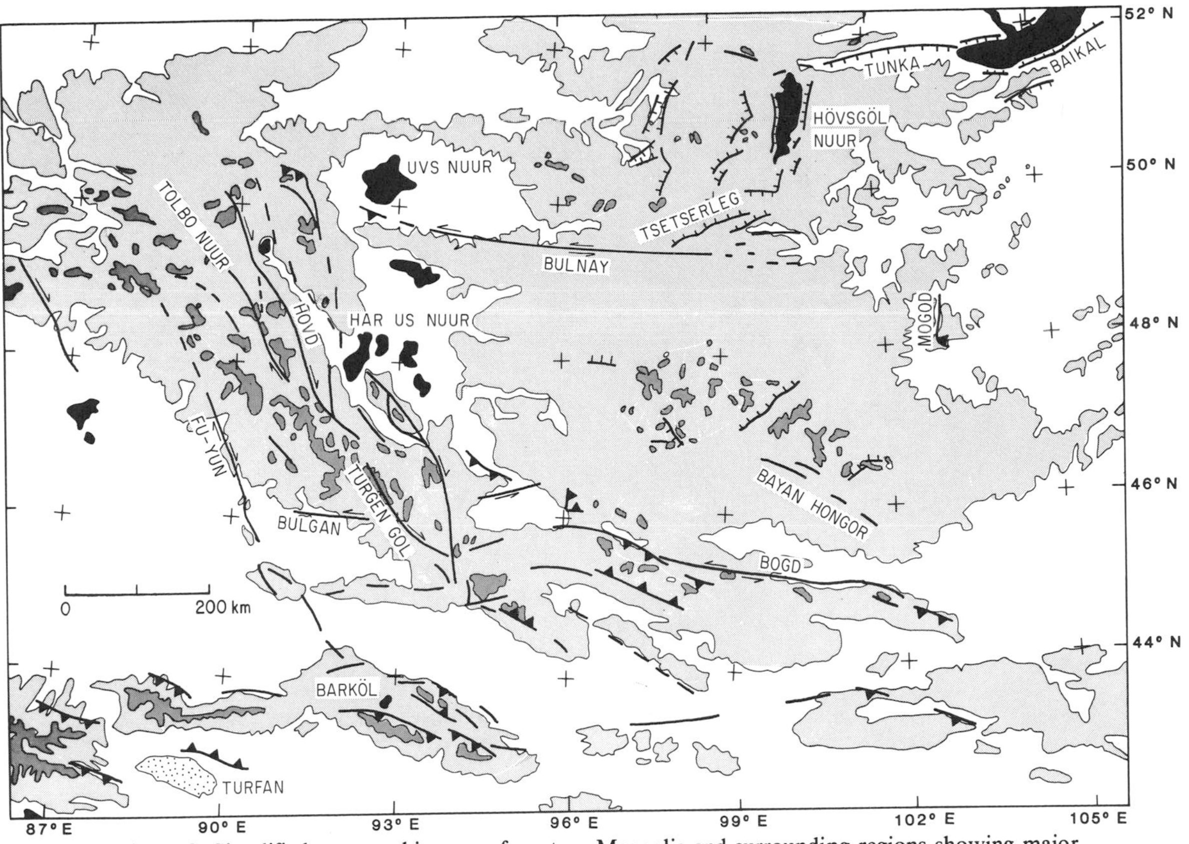

Figure 3. Simplified topographic map of western Mongolia and surrounding regions showing major active faults. Topography as in Figure 2. Most such faults were taken from Khil'ko and others (1985) and Tikhonov (1974). Dark lines with arrows parallel to them show strike-slip faults, and arrows show sense of slip. Dark lines with triangular teeth indicate thrust faults, and teeth point in downdip directions. Dark lines with hatching indicate normal faults, and hatching points in the downdip direction.

feev and Nikolaeva, 1982, p. 82), more recently than in the Mongolian Altay, as if the Gobi Altay might be at an earlier stage of development.

The Mongolian Altay and Gobi Altay are separated from another highland in the interior of western Mongolia, the Hangayn Nuruu, by the Ih Nuuryn Hotgor ("Depression of Large Lakes") and the Nuuryn Höndiy ("Valley of Lakes"), respectively (see Glossary). Devyatkin (1975, p. 271) stated that the thickness of Neogene-Quaternary sediment in the Ih Nuuryn Hotgor is not less than 500 to 700 m. Thus the possibility of very deep basins cannot be ruled out. Yet, the absence of large isostatic gravity anomalies (Zorin and others, 1982) implies that depths to basement are not greater than 1 to 2 km. Instead, the Ih Nuuryn Hotgor and Nuuryn Höndiy may simply be the sediment-filled edges of the Hangay, where it abuts against the Mongolian Altay and Gobi Altay. Active tectonics seems to be concentrated on these outer (western and southern) edges of the basins.

Unlike the Mongolian Altay and Gobi Altay, which have been built by late Cenozoic crustal shortening and thickening, the Hangay is defined by a broadly warped and elevated late Cretaceous–Paleogene erosion surface, or "vault" (e.g., Devyatkin, 1975, 1982a; Korina and Nikolaeva, 1982). Rare short normal faults constitute the primary evidence for young deformation (Zorin and others, 1982). Farther north, another broad highland surrounds the Hövsgöl graben system and continues northeast around the Baikal rift zone.

Thus, the various high terrains of western Mongolia do not seem to owe their existences to a common tectonic process. Crustal shortening seems to have built the Mongolian Altay and the Gobi Altay but not the Hangay or the region surrounding the Hövsgöl graben system.

Upper mantle structure of western Mongolia

The structure of the upper mantle beneath both the Hövsgöl graben system and the Hangayn Nuruu also seems to differ from that beneath the Mongolian Altay and the Gobi Altay. Zorin and others (1982) emphasized that the low seismic wave velocities and other evidence for high temperatures in the mantle beneath the Baikal rift system also characterize the upper mantle beneath the Hövsgöl and Hangay regions. The mean Pn velocity for western Mongolia is 7.92 (± 0.22) km/s (Baljinnyam and others,

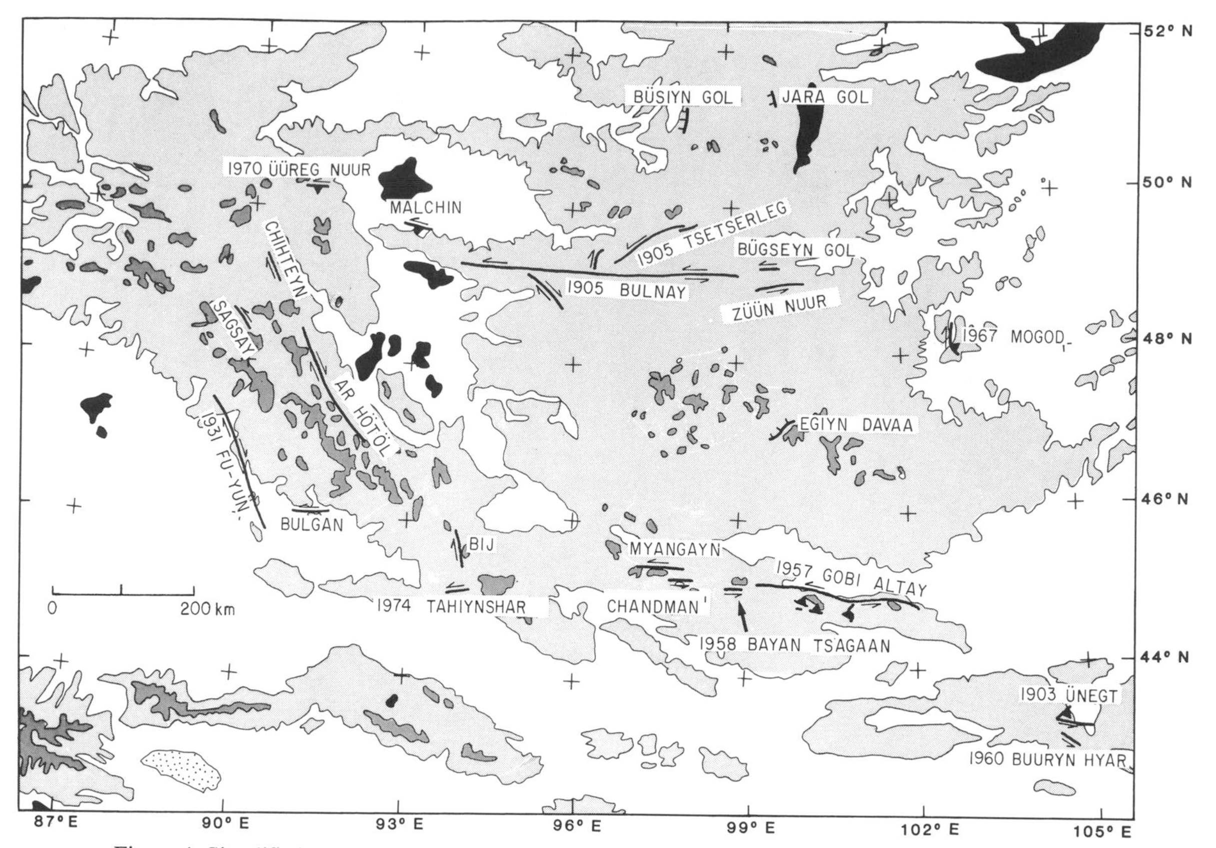

Figure 4. Simplified topographic map of western Mongolia and surrounding regions showing surface ruptures of earthquakes mapped by Khil'ko and others (1985). Topography and symbols for faulting as in Figure 3.

1975), but Zorin and others (1982) noted that studies using earthquakes in the Mongolian Altay and stations in the USSR suggest a Pn velocity for that area of 8.1 (± 0.1) km/s (Dantsig and others, 1965; Zhalkovskii and others, 1965). Lower velocities beneath the Hangay and Hövsgöl regions may pull the average down to only 7.9 km/s. In addition, P-wave residuals at stations in Siberia and Mongolia from nuclear explosions in Nevada consistently show delays not only at stations near the Baikal region but also in the Hövsgöl and Hangay regions (Rogozhina and others, 1983). Finally, late Cenozoic, including very recent, basaltic volcanism is scattered over parts of the Hövsgöl and Hangay regions but not over the Mongolian Altay or Gobi Altay (e.g., Devyatkin, 1981, p. 104–118).

Many of Khutorskoy and Yarmolyuk's (1990) 27 heat flow measurements from the central part of Mongolia can also be interpreted as being consistent with high temperature beneath the Hangay and Hövsgöl regions and lower temperatures beneath surrounding areas, but they do not refine this image. Although heat flow measurements made through sediment in Hövsgöl show a large scatter, seven of nine measurements are greater than 60 mW/m^2, and four are greater than 100 mW/m^2, compared with values closer to 40 mW/m^2 typical of shields (e.g., Sclater and others, 1981). Two measurements made just south of Hövsgöl also give roughly 60 mW/m^2. A similar large scatter with many high values characterizes heat flow measurements from the Baikal region (Lysak, 1978, 1987). In contrast, most of the lowest values in Mongolia are from the south-southeastern part of the country, where five measurements range from 16 to 40 mW/m^2.

Khutorskoy and Yarmolyuk's (1990) other nine heat flow measurements, made east of the Hangay, are higher than one would expect if the high-temperature upper mantle were confined to the Hangay-Hövsgöl region. They show a range between 31 and 100 mW/m^2, with most between 60 and 80 mW/m^2. Khutorskoy and Yarmolyuk (1990) reported that the reduced heat flow, that conducted from the underlying mantle, is only 38 mW/m^2. This range of values is somewhat greater than that for shields (Sclater and others, 1981) but is by no means large for a region that lies at a mean elevation of 1,000 m without having undergone crustal shortening since 150 Ma. Thus, a major contributor to this high heat flow would be high rates of radiogenic heating in the upper crust (upper 11 km, according to Khutorskoy and Yarmolyuk [1990]).

The interpretation of nearly all of these heat flow measurements is also complicated by the possibility that the present tectonic regime was established only within the last few million years. The measured heat flow may bear less on the present tectonic regime than on the mantle dynamics of, say, the Miocene epoch. Nevertheless, the high values from high regions concurs with the inference of a hot uppermost mantle beneath the Hangay and Hövsgöl regions.

The extent to which high elevations elsewhere in Mongolia are due to Cenozoic crustal thickening should be revealed by determinations of crustal thicknesses, but these are few, and those requiring lateral variations in thickness are fewer. Seismic refraction measurements from the west end of the Baikal rift zone indicate relatively thick crust of 45 to 50 km for an area that has not undergone crustal shortening since late Precambrian time (Mishen'kin and others, 1978). The reported thickness decreases somewhat beneath the northern end of the Hövsgöl region to 43 to 45 km, but these crustal thicknesses are larger than those of typical continental crust (35 to 40 km). Although the basement of the rest of Mongolia was consolidated since Precambrian time, it is probably sensible to assume that a comparable thickness of crust underlies most of western Mongolia (Zorin and others, 1982). More importantly, phase velocities of Rayleigh waves with periods between 10 s and 40 s and crossing western Mongolia are relatively low. Kozhevnikov and others (1990) interpreted these as evidence for an average crustal thickness of 50 km. In fact, their measured phase velocities are very similar (within 0.1 km/s) to those determined by Ewing and Press (1959) for roughly the same period range and for paths across the Basin and Range Province of the United States. Given that the crust is relatively thin (30 km) in that latter region, however, the crustal thickness of 50 km cannot be well constrained by these low Rayleigh wave phase velocities alone. For instance, the crustal thickness may be only 40 km throughout much of western Mongolia.

There is a suggestion of relatively thick crust beneath the Mongolian Altay. Deviations from isostatic equilibrium throughout Mongolia apparently are small. Assuming local Airy isostasy, Zorin and others (1982) constructed a map of hypothetical crustal thicknesses for the whole country. Their map, therefore, reflects the regional elevations, with thickest crust beneath the Mongolian Altay. Using P-wave arrival times from earthquakes in the Mongolian Altay and recorded in Siberia, Zhalkovskii and others (1965) inferred an average crustal thickness of 44 ± 6 km for this area. In an extension of this work, however, Tsibul'chik (1967) inferred values of 38 to 43 km for the foothills and 48 to 53 km for the interior of the Altay, along the border of Mongolia with the Russia.

The assumption of local isostatic equilibrium also yields a thick crust beneath the Hangay, where crustal shortening has not occurred for hundreds of millions of years. As discussed by Zorin and others (1982), a hot upper mantle seems to underlie this area. Thermal expansion of this material, instead of a thick crust, may compensate for the high elevations. Thus, a relatively uniform crustal thickness may underlie the Hangay and the area around the Hövsgöl graben system.

The important features of this deep structure are: (1) where evidence for crustal shortening is clearest (the Mongolian Altay and Gobi Altay), there is no suggestion of a warm upper mantle, and (2) where there is a suggestion of a warm upper mantle (beneath the Hangay and Hövsgöl graben system), active deformation seems to include a relatively large component of normal faulting and no major thrust faulting.

Brief summary of the Late Cretaceous and Cenozoic geologic history

Mongolia seems to have been tectonically quiet when the Mesozoic era ended. Most of the relief and active structures apparently formed in the latter part of the Cenozoic era (e.g., Devyatkin, 1974, 1975; Devyatkin and Shuvalov, 1990; Kozhevnikov and others, 1970; Tikhonov, 1974). Flat topography, commonly treated as a Cretaceous or early Cenozoic erosion surface, characterizes much of the topography of western Mongolia. In the Hangay, this surface extends over a vast area. In the Mongolian Altay, the remarkably flat tops of high mountains are ascribed to recent uplift of this surface above the surrounding territory. In parts of Ih Nuuryn Hotgor, the Nuuryn Höndiy, and the Gobi Altay, Cretaceous and Paleogene deposits, termed a "platform mantle" ("platformennyi chekhol") by Devyatkin (1974, p. 185; 1981), cap this surface and define its age. Flat basalt flows of late Oligocene or early Miocene age cover this surface in some areas [Devyatkin, 1981; Kozhevnikov and others, 1970]. Throughout most of the Mongolian Altay in western Mongolia, however, Cretaceous and Paleogene deposits seem to be absent (e.g., Dergunov and others, 1980; Devyatkin, 1981]. Hence, the ages of both the erosion surface and its present elevation are not well constrained.

Conglomerate and other coarse continental sediment (molasse) assigned a Neogene age overlie this erosion surface in some areas. The change to coarser sediment seems to be the main argument for the inference that deformation began at the end of the Oligocene or beginning of the Miocene epoch. In places, this coarse sediment overlies red argillaceous Oligocene sandstone (Devyatkin, 1974, p. 186; 1975). Elsewhere, the conglomerate is middle Oligocene (Devyatkin, 1970; Dashzeveg, 1970). In some places, early Miocene conglomerate overlies late Oligocene finer material, but elsewhere, apparently middle Miocene conglomerate overlies the Oligocene series directly (Devyatkin, 1981, p. 14, 50–57). Late Miocene deposits have been found only adjacent to the Mongolian Altay (Deyyatkin, 1981, p. 57–61; Devyatkin and Shuvalov, 1990, p. 170). Thus, the changes in sedimentation, and presumably tectonic processes that created relief, apparently varied spatially and temporally. Moreover, the timing of such presumed tectonics does not appear to be precisely determined.

Two distinct types of sediment characterize early and middle Pliocene deposits (Devyatkin, 1981, p. 15, 61; Devyatkin and Shuvalov, 1990). Lake sediment was deposited in the large basins of the Ih Nuuryn Hotgor east of the Mongolian Altay and in the Nuuryn Höndiy north of the Gobi Altay (e.g., Devyatkin, 1981, p. 65). A dominant component of gravel is present at the edges of mountainous regions (Devyatkin, 1981, p. 62–63; Devyatkin and

Shuvalov, 1990, p. 171). This sediment suggests that in early Pliocene time the present mountainous areas of western Mongolia had already been elevated and these depressions existed.

Devyatkin (1974, 1981, 1982b) and Kozhevnikov and others (1970) reported that the initial phase of Cenozoic deformation was slow and then accelerated in middle Pliocene to Pleistocene time. Again, the evidence for an acceleration includes the abrupt increase in the rate of deposition of coarse sediment, both in intermontane basins within the mountain belts and in lower basins adjacent to presently high terrains. The late Pliocene and Pleistocene deposits in basins within the Mongolian Altay include a significant fraction of glacially transported debris (Devyatkin, 1981, p. 13). As the late Pliocene epoch was a time of pronounced global climate change, however, it is possible that the geologic evidence used to infer an abrupt change in tectonic style actually reflects a change in erosion rate due to the climatic change to glacial conditions and not to an abrupt change in tectonics (Molnar and England, 1990).

Regardless of whether the initiation of Cenozoic tectonics is assigned an Oligocene or a Pliocene age, Cenozoic deformation in Mongolia postdates the beginning of collision between India and Eurasia (e.g., Molnar and Tapponnier, 1975).

THE MONGOLIAN ALTAY

This mountain belt trends northwest-southeast and is cut by an anastomosing network of north-northwest– to northwest-trending faults. Along the northeast flank, a sharp break in slope separates the range from the Ih Nuuryn Hotgor and implies recent faulting (e.g., Devyatkin, 1975, p. 270). Geomorphic evidence of several types attests to both right-lateral strike-slip and thrust or reverse components of slip within the Mongolian Altay (Figs. 2, 3, and 4). In general, however, oblique slip on individual faults appears to be minor.

The strike-slip faults are particularly clear on the Landsat Imagery (Tapponnier and Molnar, 1979). Rivers follow many of them (Devyatkin, 1974, p. 191). From observations made on the ground, Tikhonov (1974) emphasized that right-lateral slip characterizes many of these faults: in particular the Hovd, the Tolbo Nuur, and the Turgen Gol faults (Fig. 3). He stated, however, that although some stream valleys seem to be displaced, no estimates of the total amounts of slip could be given. Unfortunately, the dirth of Mesozoic and Tertiary rock within the Mongolian Altay of westernmost Mongolia (Dergunov and others, 1980) is likely to make it difficult to distinguish amounts of Cenozoic and older displacements. Trifonov (1983, p. 97) and Khil'ko and others (1985, p. 132) reported maximum stream offsets along the Hovd fault of 3.5 km and at least 6 km, respectively, but clearly larger cumulative offsets cannot be disproven easily. Stream offsets of a few kilometers are clear on the Landsat imagery of the Hovd and Tolbo Nuur faults (Fig. 5). In addition, segments of some of these strike-slip faults show evidence of very recent surface ruptures. The most recent rupture of the region is associated with the 1931 Fu-yun earthquake in China. Most of the other ruptures cannot be associated with recorded earthquakes, but one, the Ar Hötöl, seems to be related to a very large earthquake in 1761.

The surface ruptures also show that deformation is not purely right-lateral shear. Examples of conjugate left-lateral slip on east-west–trending faults are present. Evidence for thrust faulting with roughly northeast-southwest shortening is provided by geologic observations in the field (Devyatkin, 1974; Tikhonov, 1974), deformation seen on the Landsat imagery (Tapponnier and Molnar, 1979), and fault plane solutions (Fig. 2). Reverse faulting is particularly clear along the northeastern margin of the Mongolian Altay. For instance, Devyatkin (1974, p. 188; 1975, p. 266) described a thrust fault about 150 km long along the northeastern edge of the Zereg basin (Fig. 6). Paleozoic rock in the range to the east (the Züün Jargalantyn Nuruu) has been thrust southwest onto Jurassic rock, and the Jurassic sequence has been thrust onto Cenozoic rock in the basin along a fault dipping 25 to 30° northeast. Surface ruptures showing reverse faulting in the Mongolian Altay, however, are much less abundant (or apparent) than those showing strike-slip faulting. We are aware of only relatively minor, clear, recently active thrust faults in the Mongolian Altay.

Right-lateral strike-slip faulting parallel to the Mongolian Altay

Let us begin with a discussion of right-lateral faulting and then consider the conjugate left-lateral before discussing examples of thrust or reverse faulting.

Fu-yun fault. The Fu-yun earthquake, of 1931 August 10 (46.89°N, 90.06°E, M = 8 [Ding, 1989; Gu and others, 1989] and M = 7.9 [Richter, 1958]), occurred in Xinjiang, China, just west of Mongolia, and ruptured the fault of the same name. The fault that ruptured in the Fu-yun earthquake is clear on the Landsat imagery and was recognized without any knowledge of the 1931 surface rupture (Tapponnier and Molnar, 1979). The mapped surface rupture trends north-northwest (340°) for a distance of 180 km. Right-lateral displacements were observed along most of the fault (Fig. 7) (Shi and others, 1984; Yang and Ge, 1980; Zhang Pei-zhen, 1982). Several measurements of 9 to 11 m were made in the central section, with an isolated maximum measurement of 14.6 m. Measured displacements decrease monotonically to both the north and the south (Fig. 7), with an average of about 8 m. Components of normal faulting in the northern part of the rupture and of reverse faulting in the southern part were reported, implying scissors faulting and suggesting that a small component of relative rotation about a vertical axis might have occurred. Vertical components of 1 to 3.6 m were noted, but they were generally smaller than the strike-slip components in the same localities. The measured displacements include deformation associated with aftershocks, one of which (1931 August 18) was assigned a magnitude of 7.2.

An average slip of 8 ($\pm$ 2) m along a zone 180 km in length, extending to a depth of 20 km, yields a scalar seismic moment of 9.7 ($\pm$ 2.4) $\times 10^{20}$ N m (Molnar and Deng, 1984), indistinguish-

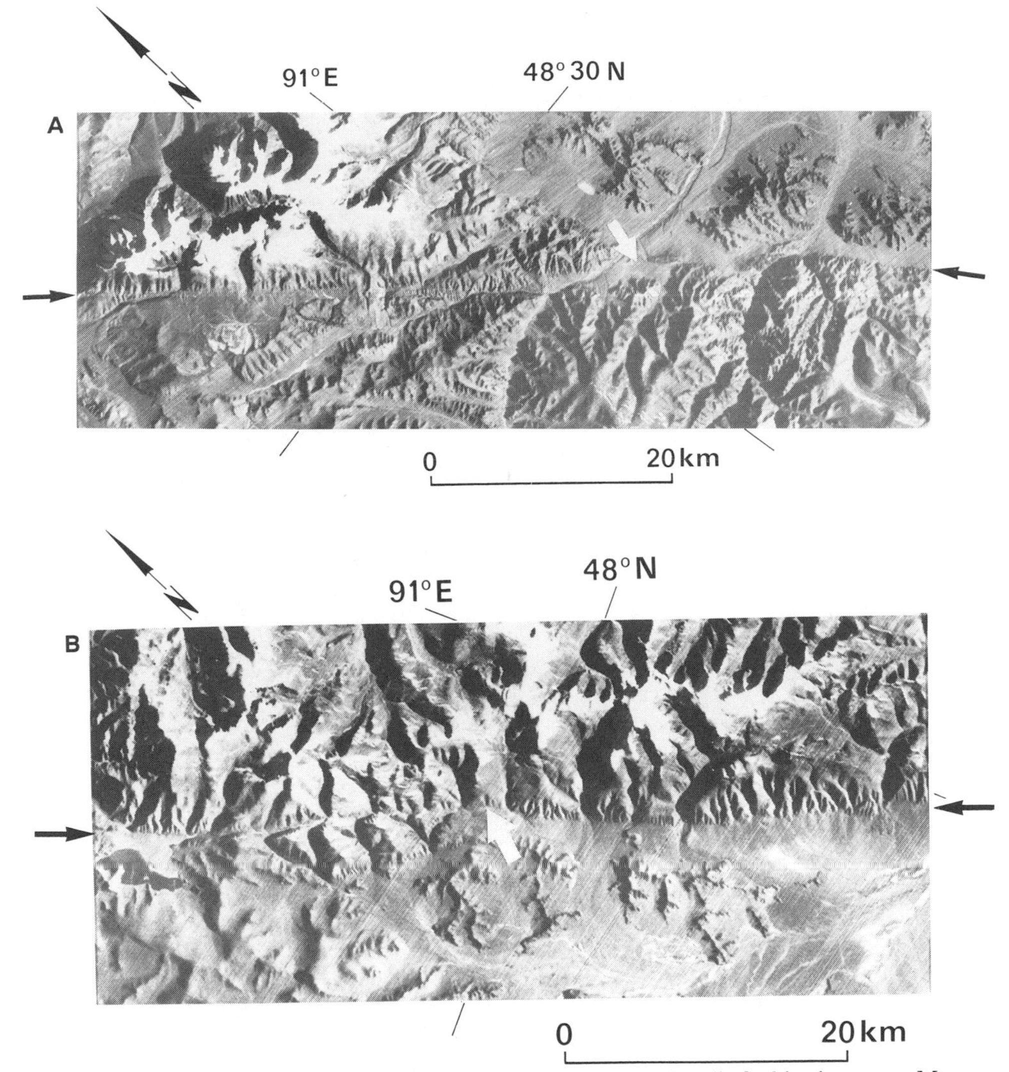

Figure 5. Portions of Landsat image 1506-04200-5 showing active strike-slip faulting in western Mongolia. a, The black arrows on the sides and the white arrow point to the Hovd fault, including the northern end of the Ar Hötöl rupture zone. Notice the deeply dissected, high (snowcapped), but otherwise flat, terrain in the upper left part of the image. The basin farther north is the Achit Nuur basin. b, The black arrows on the sides and the white arrow point to the Tolbo Nuur fault. The left edge of this segment is roughly 30 km southeast of the lake Tolbo Nuur. Notice the clear left-lateral offsets of drainage on the left side of the image.

able from the value inferred from amplitudes of long-period surface waves: 8.5×10^{20} N m [Chen and Molnar, 1977].

Shi and others (1984, p. 336) reported offsets of some features of 20 m and assigned them to cumulative slip associated with previous earthquakes. Using Carbon-14 ages of material near offset features, they deduced a recurrence interval of 230 yr, but they did not describe completely the facts on which that deduction was based. Streams flowing west from the higher parts of the Altay range are displaced right-laterally 100 m to 3 km at the Fu-yun fault, with a number of them clearly offset 1.5 to 2 km (Fig. 8). From the recurrence interval and from a mid-Pleistocene age assigned to these offsets, Shi and others (1984, p. 339) inferred a slip rate of 10 mm/a on the Fu-yun fault. We cannot evaluate the ages that they assigned, but the offsets, such as those shown in Figure 8, are clear on maps that they and Zhang Peizhen (1982) presented. Thus, even if the rapid stream incision dated from late Pliocene time (2.5 Ma) instead of middle Pleistocene time, the slip rate would be at least 1 mm/a.

The Fu-yun fault seems to slice obliquely across the northwesterly trending Paleozoic Altay orogenic belt. Geologic mapping, which we cannot evaluate, by Chinese scientists suggests that Paleozoic faults, sedimentary units, and granites have been

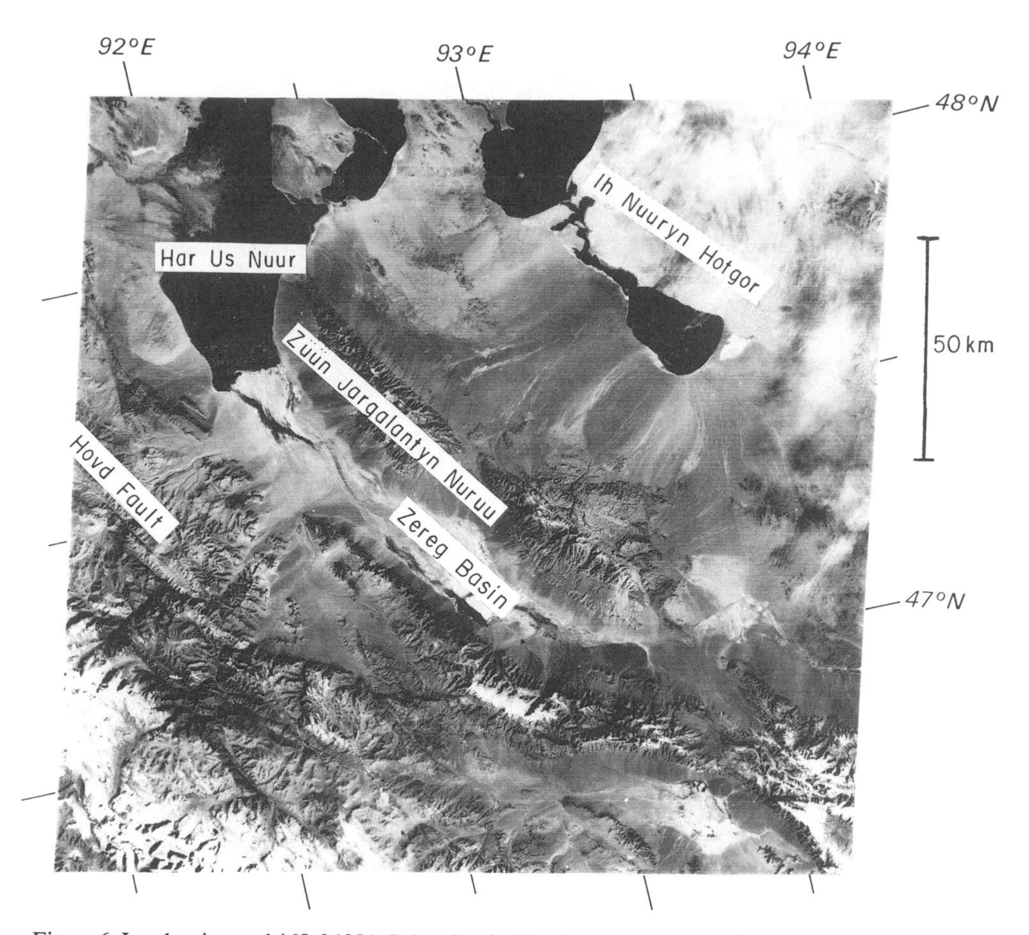

Figure 6. Landsat image 1468-04091-5 showing faulting in western Mongolia. The lake Har Us Nuur is in the northwest part of the image, and a clear range of mountains, the Züün Jargalantyn Nuruu, trends south-southeast from the lake. Faults bound the range on both its east and west sides. Those on the east may include large strike-slip components. Those on the west are characterized by large thrust or reverse components. The Zereg basin lies southwest of this range, and a clear scarp bounds its southwest side also. The Hovd fault enters the image from the middle of the west side, near 48.2°N, and is marked by a linear trace.

offset 25 to 30 km (Fig. 9) (e.g., Ding, 1989; Shi and others, 1984; Zhang Pei-zhen, 1982). An offset of this amount would accumulate in only a few million years, if the slip rate were several mm/a.

Sagsay fault (Figs. 4 and 10). Khil'ko and others (1985, p. 40–41) reported a surface rupture trending 330 to 325° for a distance of 35 to 37 km near 48.5°N, 89.7°E (Fig. 10). Both tension gashes, 40 to 50 m long, 2 m deep, and trending 350 to 010°, and mole tracks, 2 to 3.5 m high and trending 100 to 120°, attest to right-lateral slip. Khil'ko and others (1985) inferred a maximum strike-slip offset of "not less than 3 m" and a mean of 2.5 m, with a maximum vertical offset of 2 m and a mean of only 0.5 m. The displacement includes a vertical component, but the amount and sense vary along the rupture, as is common with overall strike-slip displacement. In our cursory examination of this rupture, we saw only one locality where it was possible to estimate a horizontal offset. In the segment northwest of Ezerleg Uul (Fig. 10), the scarp faces west and is only 0.2 m high, and a dry gully sloping west is displaced roughly 3 m right-laterally.

Khik'ko and others (1985) noted that the well-preserved scarp and steep sides of tension cracks are comparable with those of the Ar Hötöl and Bulnay ruptures described below. From this comparison, they assigned an age of 200 to 300 yr to the Sagsay rupture. The area traversed by the Sagsay rupture is above 2,000 m (Fig. 10), and both the scale and the preservation of the surface deformation may be due, at least in part, to frozen ground. We discuss the logic of this possibility below with reference to the Bulnay rupture.

The recent scarp does not follow the regional topography. In places, it lies along the west side of high terrain, but along much of its length it obliquely crosses high alluvial fans and wide gently sloping surfaces (Fig. 11). Thus, slip on this fault trace does not seem to have played an important role in shaping the regional topography. Tikhonov (1974) showed this as a minor fault on his

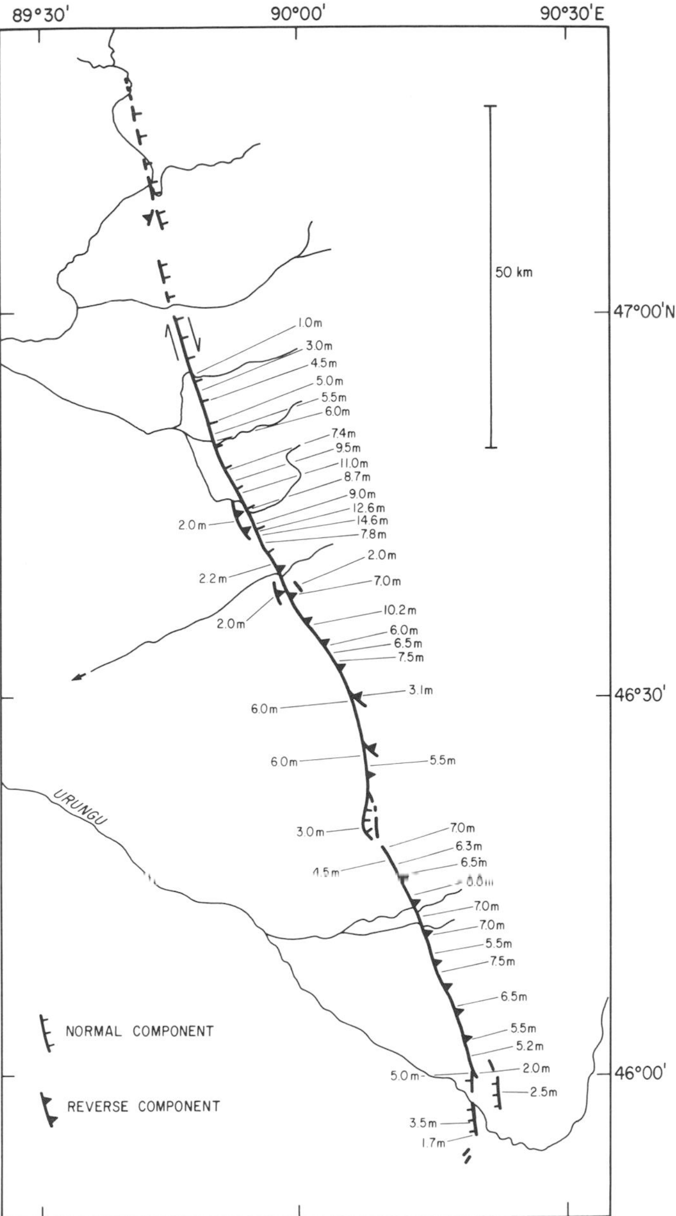

Figure 7. Map of the 1931 Fu-yun earthquake rupture zone in northern Xinjiang, showing measured right-lateral strike-slip displacements. These measured values were taken from maps by Zhang (1982) and Shi and others (1984).

map and did not discuss it at all; it is not prominent on the Landsat imagery. Apparently, slip on the Sagsay fault has not contributed much to Cenozoic regional deformation.

Tolbo Nuur fault (Figs. 3 and 5b). Khil'ko and others (1985) and Tikhonov (1974) distinguished the Tolbo Nuur fault as a major strike-slip fault between the Sagsay and Hovd faults (Figs. 3 and 4). The Tolbo Nuur fault is very clear on the satellite imagery (Fig. 5b). Right-lateral slip is revealed by offset valleys.

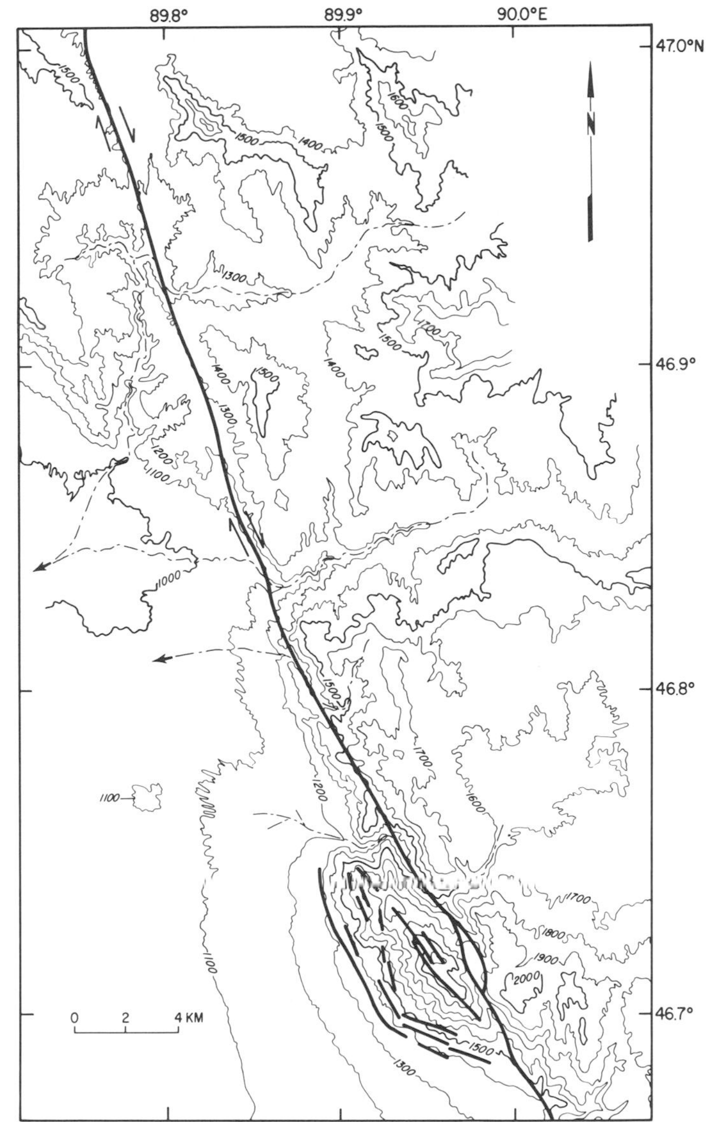

Figure 8. Topographic map showing stream courses flowing southwest across and offset by the Fu-yun fault. Offsets of 1 to 3 km are clear. (Redrawn from Zhang [1982].)

This fault, like all strike-slip faults in Mongolia, is not straight. In places, it separates high summits from long narrow valleys. Tolbo Nuur itself is a long northwesterly trending lake with a high flat ridge to its northeast. The Tolbo Nuur fault manifests itself as a clear scarp at the foot of the range. Alluvial fans are disrupted and show a component of vertical displacement (Fig. 12). We suspect that the strike-slip component is large, but we have not studied it.

Hovd fault (Fig. 3). Khil'ko and others (1985) reported two recent surface ruptures along this fault. The first, the Chihteyn segment, at 49.2°N, 90.3°E, is about 27 km long (Fig. 4). Khil'ko and others (1985, p. 25–27) described a west-facing scarp varying in height from 0.5 m to 1.5 or 2.0 m along the central 10 km of

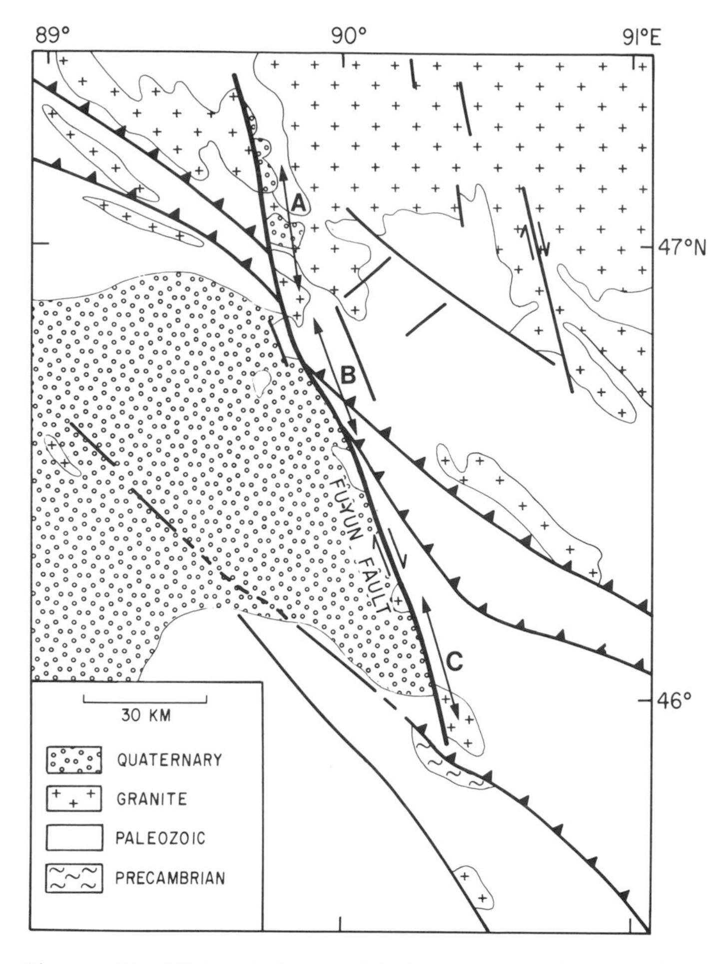

Figure 9. Simplified geologic map of the Fu-yun region showing Paleozoic faults and granite bodies apparently offset by slip on the Fu-yun fault. Granite bodies seem to be offset 35 km at A and C, and Paleozoic faults seem to be offset 35 km at B. (From Zhang [1982].)

the segment. The sense of slip is opposite to that of the regional topography; the Hovd fault in this area follows the west side of a wide basin, the Achit Nuur basin (Fig. 5a), and separates it from a high range west of the fault. Apparently the most recent vertical component of slip is opposite to that that built the present regional topography. The height of the scarp decreases farther north, where tension cracks and mole tracks are the dominant manifestations of rupture. Along much of the rupture, en echelon tension cracks with an azimuth of 010 to 030° mark the surface trace and attest to a significant component of right-lateral slip. Khil'ko and others (1985) inferred a greater strike-slip than dip-slip component. From the morphology of the features along the trace, compared with those of other ruptures in the Mongolian Altay, they assigned an age of 1,000 to 1,500 yr.

Khil'ko and others (1985, p. 41–45) mapped a second surface rupture trending on average about 330° through the low pass, Ar Hötöl, at 47.5°N, 91.8°E, west of the town of Hovd (Figs. 4 and 13). They noted that the scale of deformation and the morphology of the surface rupture are comparable with those associated with the great Bulnay earthquake of 1905 July 23, discussed below. The deformation along a zone more than 215 km in length includes very large tension cracks—with lengths of 40 to 50 m, widths up to 8 to 10 m, depths up to 2.5 to 3 m, and trends between 350 and 010°—and correspondingly large mole tracks, up to 15 to 20 m long, 2 to 3 m high, and trends of 100 to 120° (Fig. 14). These features are consistent with the large components of right-lateral slip observed at several localities. Khil'ko and others (1985) estimated the average displacement to be 4.5 m, with a maximum of 7 m. In contrast, Trifonov (1988) reported an average of only 3 m, with a maximum of 5 m.

The southern end of the rupture is marked by a series of ruptures, roughly 60 km in total length (Fig. 13). Khil'ko and others (1985) reported vertical components of 1 m to as much as 3 m and horizontal components of 1.5 to 2 m. Between this series of ruptures and the Dund Tsenher Gol (Fig. 13), however, the scarp crosses especially rugged topography and is not very clear.

The most impressive offsets lie between the Dund Tsenher Gol and Buyant Gol (Fig. 13), where there are several examples of right-lateral offsets of shallow streambeds and dry streambeds reaching 4 to 5 m. In many segments, a vertical component is clear. Khik'ko and others (1985) reported vertical components as high as 3 m. Because the scarps associated with vertical components face both east and west, they appear to reflect only local phenomena, possibly due to horizontal slip. The average sense of slip is nearly pure strike-slip.

Impressive surface deformation lies just north of the Dund Tsenher Gol, where the scarp follows the western slope of a ridge. Slip along a segment a few kilometers in length includes both right-lateral and vertical components with the west side up (Figs. 15 and 16). Where the fault trace trends parallel to the local contours, a vertical component of roughly 2 m can be measured, but where the trace crosses northerly or northeasterly trending contours, both the right-lateral and the vertical components appear to be especially large (Fig. 15). In such areas, where the slope of the topography dips northwest, the vertical component contributes to the apparent right-lateral offset, and the right-lateral component contributes to the apparent vertical offset. Using the vertical component of about 2 m measured nearby to correct the apparent vertical component, we estimated the right-lateral component of slip to be 4 to 5 m (Fig. 15). Where the north-northwesterly trending trace crosses northwesterly trending contours, however, the vertical component introduces an apparent left-lateral component, which reduces the apparent right-lateral component. In these segments, the fault trace is not clear.

In this same segment just north of the Dund Tsenher Gol,

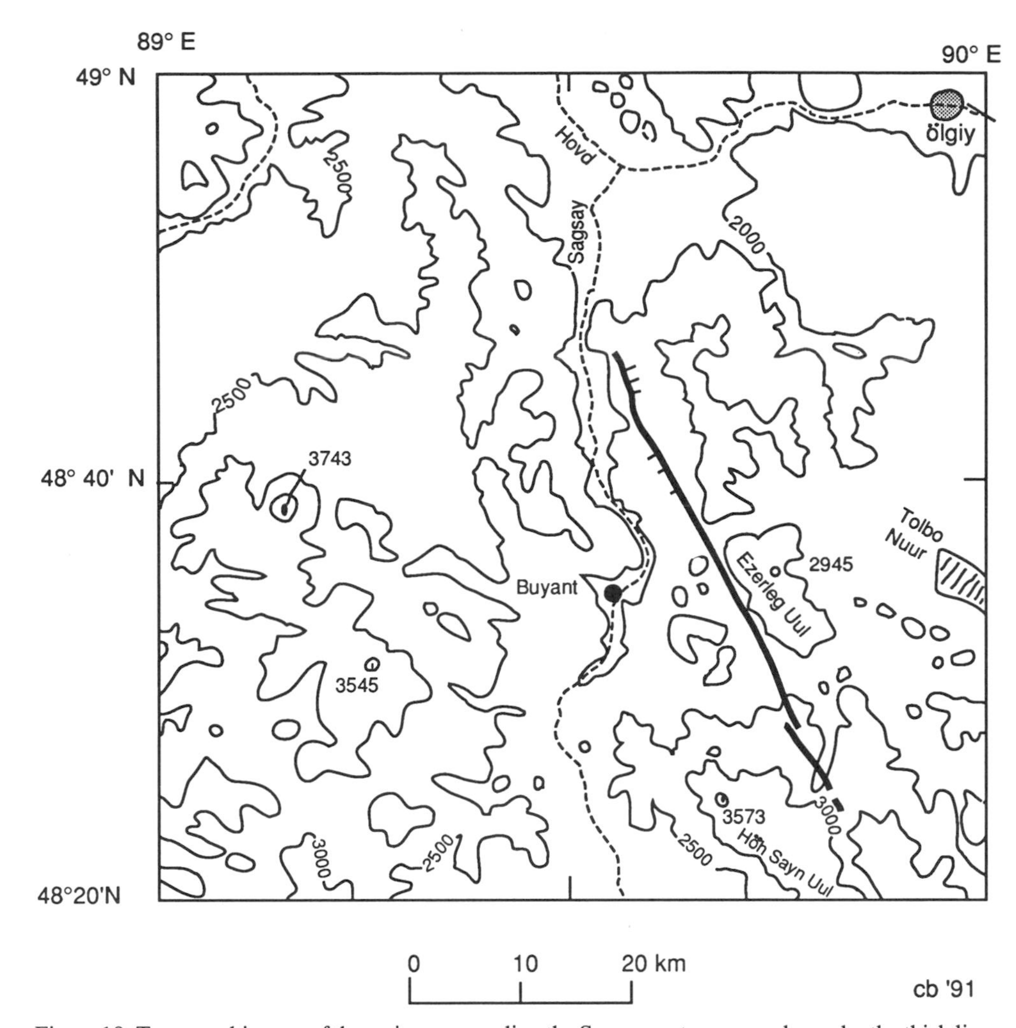

Figure 10. Topographic map of the region surrounding the Sagsay rupture zone, shown by the thick line. Note that the rupture crosses a region higher than 2,000 m. It follows the base of the ridge defined by the closed contour north of that surrounding Ezerleg Uul but for the most part crosses gentle topography. Thin dashed lines show rivers (Hovd and Sagsay). The photo in Figure 11 was taken in the area between the Ezerleg Uul and the ridge to the north.

Figure 11. Photograph taken standing on a large mole track and looking north-northwest along the Sagsay rupture (middle of photo) from the road (foreground) between Tolbo Nuur and the Sagsay River. The ridge in the background lies just north of the Ezerleg Uul (Fig. 10). Note the large tension cracks and mole tracks with relief of 1 to 2 m. (Photo by P. Molnar, August 1991.)

Figure 12. Photograph of the Tolbo Nuur fault zone, just northeast of Tolbo Nuur. View is northeast toward the base of a ridge parallel to the fault. Note the clear ruptures, with black arrows beneath them, across alluvial fans. Streams debouch from the ridge onto the basin occupied by the lake. (Photo by P. Molnar, with 135-mm lens, August 1991.)

the fault trace crosses two steep valleys. Views of the trace along its strike (Fig. 16) allow the dip at the surface to be measured. In both localities we estimated eastward dips of 52°, but this surprisingly gentle dip almost surely does not extend to great depth. Farther north, the fault obliquely crosses a pass, and yet farther it follows the east side of a ridge. Although the fault scarp along this segment is much less impressive than shown in Figure 15 and 16, the apparent dip is westward. We cannot determine the extent to which the apparent dip is due to downslope creep, but the apparently planar surface of the fault, such as that shown in Figure 16, suggests that if downslope creep is important, it is not confined to the top few meters of the surface.

In the 20-km zone surrounding Ar Hötöl, the scarp follows a well-defined linear valley or "strike-slip rift" (Fig. 14) that seems to offset one major valley 5 km (Fig. 13).

Some of the most impressive deformation is near Ar Hötöl itself (Fig. 14). Roughly 6 km south of Ar Hötöl, the scarp lies a little above the base of a steep slope, and right-lateral offsets are clear, especially from a distance. The complex topography consisting of scree and bouldery slopes, however, makes measuring an offset difficult. In one place we estimated 7.5 ± 3 m. Farther north, remarkably well preserved deep tension gashes, as deep as 2 to 3 m, characterize the zone beginning 3.5 km south of the pass. Along much of the 5-km segment crossing the pass, a vertical component of slip is difficult to discern within the broad zone of deformation. Both roughly 1 km north and south of the pass, the west side seems to have been uplifted 1 to 2 m with respect to the east side (Fig. 14). This is clearest near where the fault trace crosses ancient Turkic graves, defined by circles of stones some 20 m in diameter, 1.1 km south of Ar Hötöl. The two clearest right-lateral offsets of the stones are 5.5 (± 1) m and 3 m, but the latter, smaller offset can be seen to be only a lower bound on the slip at the edge of this grave. Trifonov (1988) also reported an offset of 3 m of the graves. Between about 500 m and 1 km north of Ar Hötöl, the scarp follows the steep east slope of a ridge across which stream gullies have been offset 5.4 (± 1) m and 5.2 (± 1) m. In the 10 km farther north from Ar Hötöl, we saw no clearly measurable offsets. According to Khil'ko and others (1985), however, the scarp is clear farther north and closer to the Buyant Gol.

Khil'ko and others (1985, Fig. 3.15) showed a subsidiary rupture roughly 5 km in length and trending northwest (305°), west of the main rupture (Fig. 13). We could see no evidence for strike-slip offsets, but the fresh northeast-facing scarp suggests a vertical component of slip of 1 to 1.5 m (Fig. 17).

In the 20 km north of the Buyant Gol, we saw no clear evidence of a recent rupture, but in the segment of the Hovd fault farther north, we did see clear scarps (Figs. 18, 19, and 20). In some places (e.g., 48°02′N, 91°22′E), the gentle slope of the scarp (Fig. 18) suggests that it might be older than the steeper scarps found south of the Guyant Gol. The clearest scarp north of the Buyant Gol (Figs. 19 and 20), at 48°05′N, 91°20′E, however, does appear to be as fresh as those farther south. At this locality, water flowing northeasterly in one minor ephemeral streambed has been dammed at the fault by an uplift of the east side (Fig. 19). The middle of the upstream reach lies 7 ± 2 m north of the downstream reach. This is the locality where Khil'ko and others (1985, Fig. 3.14) indicated the largest right-lateral offset of 7 m. Because of the clear vertical component and the damming of the upstream reach, however, we cannot be confident that the 7-m separation represents strike slip in one event. The next dry streambed farther north also is clearly offset (Fig. 20). On the flat area just to the north, a vertical component of about 0.5 m is clear (Fig. 20b). The apparent right-lateral separation of 5.5 m of the north (left) bank (Fig. 20a) suggests a component of strike slip of 4 m in this area. Scattered examples of surface deformation can be seen northward until roughly 48.2°N, 91.25°E. Although a sharp break in slope is present farther north, as far as the Shurgyn Gol, a recent surface rupture was not clear to us.

Khil'ko and others (1985, p. 45–47) pointed out that a major earthquake was felt over a large part of western Mongolia and southern Siberia on 1761 December 9, with intensities comparable to those associated with the great earthquakes of 1905, 1931, and 1957. Consequently, they tentatively assigned the Ar Hötöl rupture to that earthquake. This date, however, is not supported by all of the evidence that they presented. They re-

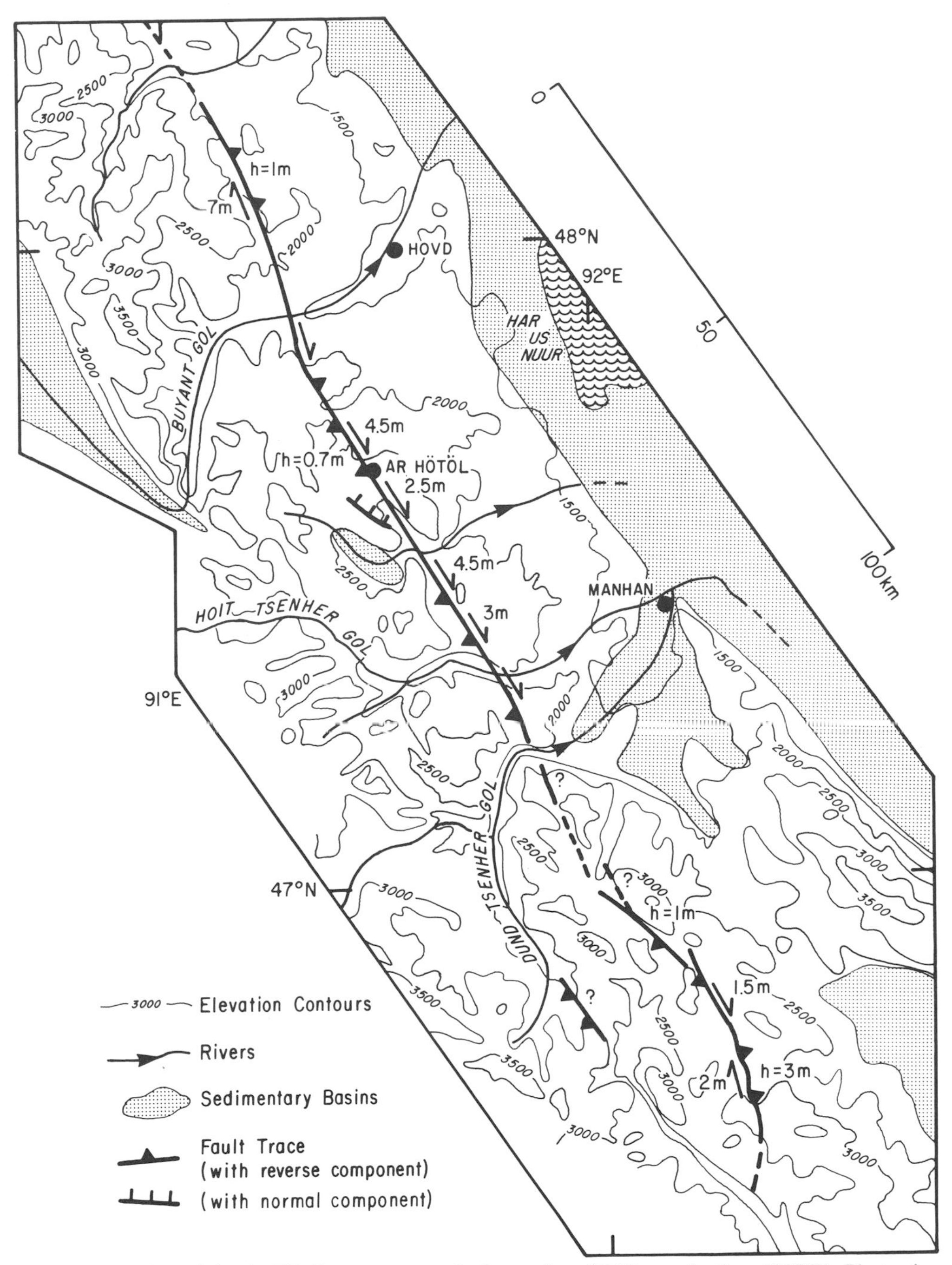

Figure 13. Map of the Ar Hötöl rupture zone (redrawn from Khil'ko and others [1985]). Photos in Figures 14, 17, 21, and 22 were taken between Ar Hötöl and the river valley to the south. Those in Figures 15 and 16 are from just north of the Dund Tsenher Gol. Those in Figures 19 and 20 were taken north of the Buyant Gol, near the "7 m" and "h = 1 m" offsets.

ported archaeological inferences that the Turkic graves near Ar Hötöl probably date from the sixth to eighth centuries and possibly from as recently as the tenth to twelfth centuries (Khil'ko and others, 1985, p. 44–45). The town of Hovd was constructed in 1685, and no mention of strong earthquakes before 1931 has been found in the town records. The absence of records of the 1905 earthquake in Hovd, however, may suggest that the older town records are incomplete. Khil'ko and others (1985, p. 45) stated that "the period from the time of the emperor Chingis Khan [ca. 1200] to the end of the fifteenth century is characterized by

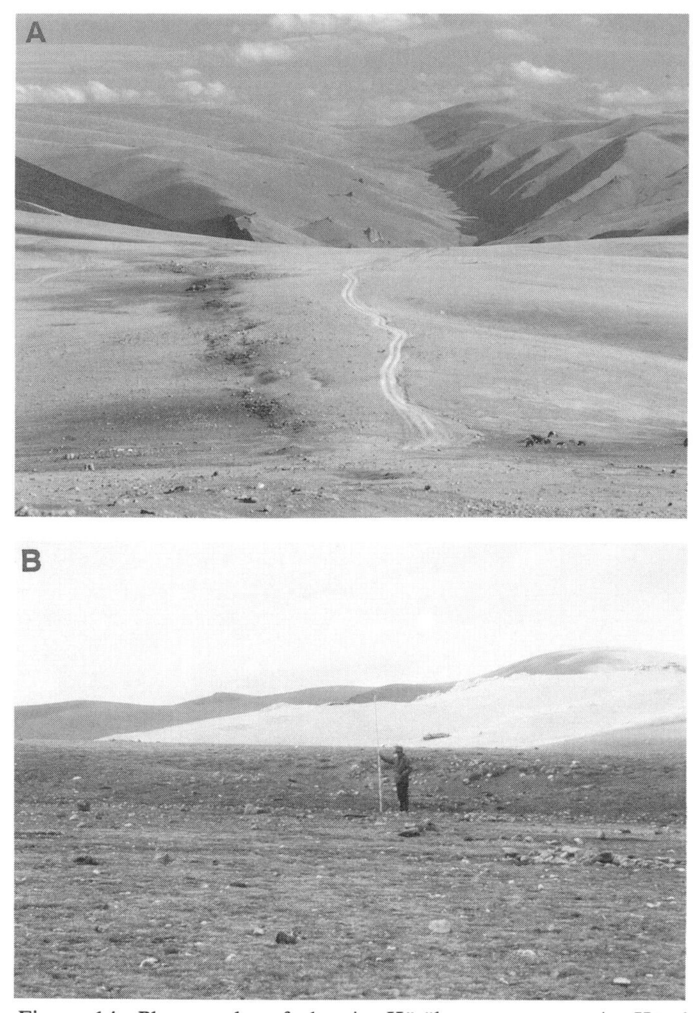

Figure 14. Photographs of the Ar Hötöl rupture near Ar Hötöl (Fig. 13). a, View southward toward the next pass, Tsas Davaa, on the horizon and roughly 35 km away. Note the large deformation in the foreground, left of the road and cattle. The west (right) side stands 1 to 1.5 m higher than the east side (see Fig. 14b). The relatively wide zone of deformation includes tension gashes and mole tracks with relief of a few meters. The scarp is also clear at the break in slope at the base of the mountain on the horizon adjacent to Tsas Davaa. The valley closer than Tsas Davaa is offset right-laterally about 5 km at the fault (Fig. 13). b, View west-southwest across area in foreground of Figure 14a. A. B. stands in front of the scarp, 1 to 1.5 m high, and holds a 3-m stadia rod. (Photos by P. Molnar, August 1991.)

Figure 15. Photographs of the Ar Hötöl rupture a few kilometers north of the Dund Tsenher (Fig. 13). All four photos show large apparent vertical and right-lateral offsets. a, View south toward the Dund Tsenher valley. The scarp is clear in the center of the photo. The photos in b, c, and d were taken looking west from the offset gully on the ridge on the left. In each of b, c, and d, A. B.—approximately 1.75 m tall—provides a scale. A vertical component approximately $\Delta u_v = 2$ m is clear roughly 30 m north of this offset gully, where the fault strikes parallel to the contours. In b, a strike-slip offset of the valley in the foreground is clear and reaches approximately 4 to 5 m. (Black arrow points at A. B.) This value can also be inferred crudely from the differences between apparent strike-slip and vertical offsets in c and d. In c, where the slope parallel to the fault is relatively steep ($\alpha = 21°$, see Fig. 37), the apparent vertical component is approximately 4.15 m. In d, where that slope is only $\alpha = 10°$, the apparent vertical component is approximately 3.5 m. Because the apparent vertical component is given by $\Delta u_v + \Delta u_{ss} \tan \alpha$, the values estimated from c and d are $\Delta u_v = 2.9$ m and $\Delta u_{ss} = 3.3$ m. (Photos by P. Molnar, August, 1991.)

relatively complete annals," but no catastrophe of the kind represented by the surface rupture is mentioned for any of Mongolia. Khil'ko and others (1985) recognized that this evidence would suggest that the earthquake occurred either between roughly the sixth and twelfth centuries or between the fifteenth century and 1685, but from the freshness of the scarps and the occurrence of

Figure 16. Photos of the Ar Hötöl rupture just north of the Dund Tsenher, approximately 750 m south of those in Figure 15. a, View looking south-southeast along the rupture. A clear eastward dip of about 52° can be seen. A black arrow in the center of the photo points at A. B., who provides a scale in both photos. b, View of the same locality, looking west-southwest across the fault trace. (Photos by P. Molnar, August 1991.)

Figure 17. Photo of a subsidiary rupture a few kilometers west of the Ar Hötöl rupture, a few kilometers southwest of Ar Hötöl (Fig. 13) and about 2 km west of the foreground in Figure 14a. View is north-northwest along the rupture and shows the break in slope at the scarp. A. B. holds a stadia rod 3 m high; vertical component is 1.2 m. (Photo by P. Molnar with 135-mm lens, August 1991.)

Figure 18. Photo of a scarp along the Ar Hötöl rupture north of the Buyant Gol, near 48°02′N, 91°22′E (Fig. 13). View is north-northeast along the rupture. The break in slope between I. B. (left) and M. G. D. (right) marks the scarp, roughly 1.5 m high. (Photo by P. Molnar, August 1991.)

Figure 19. Photo of an offset along the Ar Hötöl rupture north of the Buyant Gol, near 48°05′N, 91°20′E (Fig. 13). View is east-northeast, toward the scarp, where a dry streambed has been dammed. A. B. stands at the eastern end of ponded sand and gravel, abutting a hill on the east side of the rupture. This dry streambed is deflected to the right and then passes farther east past the hill behind A. B. The continuation of the stream has cut a deep channel, and interpreting the measured distance of 7.5 m from the center of the ponded gravel to fresh gully is not straightforward. (Photo by P. Molnar, August 1991.)

Figure 20. Photos of an offset along the Ar Hötöl rupture north of the Buyant Gol, near 48°05′N, 91°20′E and about 50 m north of that in Figure 19. Views are northeast, toward the scarp. a, The left bank of a dry streambed has been offset right-laterally. The apparent offset between I. B. (left) and A. B. (right) is about 5.5 m. Part of the apparent offset, however, is due to a vertical component. b, View northeast toward the scarp, approximately 10 m north of that in a, shows a vertical component of approximately 0.5 m (between black arrows). I. B. stands approximately 1.80 m tall and provides a scale. By logic similar to that used for Figure 15, with $\alpha = 20°$, 1.4 m of the apparent strike-slip component would be due to the vertical component. Hence, the estimated strike-slip component is approximately 4 m. (Photos by P. Molnar, August 1991.)

the earthquake in 1761, they inferred an age of only 200 to 300 yr. In contrast, Trifonov (1985; 1988) assigned an older date, before the radiocarbon age, 460 ± 100 B.P., of material that he found in two excavations and that he assumed to have been deposited after the time of the most recent surface deformation. As he gave little information making it clear why deposition of this material postdated the earthquake, it seems reasonable to us to assign the Ar Hötöl rupture to the 1761 earthquake.

Trifonov (1985) also stated that offsets of gullies and ridges near Ar Hötöl seem to cluster in multiples of 4.5 m. Accordingly, he concluded that slip has occurred in rare earthquakes with such average displacements. By using an older radiocarbon age of 1190 ± 80 yr from material below that with an age of 460 yr, he deduced that the previous earthquake occurred roughly 700 to 750 yr before the most recent earthquake. Assuming 4.5 m of slip every 700 to 750 yr, he offered an average slip rate of 6 mm/a. From his brief descriptions, we cannot evaluate the uncertainties in these deductions. Because we did not see clear offsets that were integral multiples of 4.5 m, we are unconvinced of his inference of a 700- to 750-yr recurrence interval or a 6 mm/a slip rate.

In our reconnaissance visit in 1991, we saw a few late Quaternary offsets that clearly cannot be assigned to one earthquake. Two of them are likely to have formed since the last glacial maximum. At 48°05′N, 91°20′E, where water flowing intermittently in an ephemeral streambed has been dammed (Fig. 19), a projection of the downstream course lies 28 ± 10 m south of the upstream course. The large uncertainty reflects the difficulty of choosing a thalweg for the lower reach. Approximately 4 km south of Ar Hötöl, a stream flowing west-southwest through a steep valley crosses the fault and has incised a valley about 3 to 4 m into the flatter landscape west of the fault (Fig. 21). The offset of this valley also is difficult to measure, but we estimated a right-lateral separation of the northern edge of 45 ± 15 m. The sharp incision of the lower reach of the former stream and of both reaches of the latter one may be due to increased erosion resulting from late Pleistocene climate change. The logical period to which to assign that change is the warm period, at roughly 12 ka, after the last major glacial maximum. If such an inference were correct, it would suggest an average slip rate of 2 to 4 mm/a.

We also saw two clearer but larger offsets. One is for an older course of the stream 4 km south of Ar Hötöl (Fig. 21). A wide valley, north of the present stream, emanates from near the fault zone but contains no stream or evidence for significant recent flow. It seems to be an abandoned valley. Its northern margin lies 160 (± 20) m north of the northern margin of the upstream reach, suggesting as much right-lateral offset before being abandoned. If the late Quaternary slip rate were 2 to 4 mm/a, this offset would then record displacement since about 40 to 80 ka. A few hundred meters north of Ar Hötöl, a wide (roughly 100 m) dry valley with a relatively steep gradient, with a bottom consisting of angular cobbles, and with steep sides is offset right-laterally roughly 75 m (Fig. 22). We measured offsets of 73 m for the southern edge and 74 m for the northern edge, with an uncertainty of 10 m for each. If the slip rate were about 2 to 4 mm/a, this valley would have been initiated during or at the end of the last glacial period, which culminated at roughly 20 ka (Bard and others, 1990). The shape of the valley permits a glacial origin, but this observation hardly constitutes support for a slip rate of 2 to 4 mm/a. These offset valleys taken together suggest that slip occurs more rapidly than 1 mm/a, but more slowly than 10 mm/a.

Figure 22. Offset river valley just north of Ar Hötöl. View is east, obliquely across the fault. The recent rupture can be seen in the upper right, at Ar Hötöl. The scarp and road seen in Figure 14a continue farther to the upper right. The boulder-strewn valley in the foreground, with a flat cross-sectional floor, slopes steeply east-northeast and is displaced 70 to 75 m at the Ar Hötöl fault. (Photo by P. Molnar, August 1991.)

Figure 21. Offset river valley 4.1 km south of Ar Hötöl (Fig. 13). View is east-northeast, and the Hovd fault crosses the area near the middle of the photo. A stream is incised about 3 to 4 m into the landscape east of the fault and debouches into a wider, gravelly floor at the fault. The northern bank is offset right-laterally about 45 (± 15) m at the fault. To the left (north) of this valley, another wide valley (lower left center of photo) appears to be beheaded by slip on the fault. It lies abandoned by the river that cut it and 162 ± 20 m north (left) of the present stream valley. (Photo by P. Molnar, August 1991.)

Khil'ko and others (1985, p. 32–33) reported one other right-lateral surface rupture, the Bij rupture (Fig. 4), farther south along the Tonhil fault near 45.7°N, 94.1°E (Fig. 3). Right-lateral faulting characterizes a zone 25 km long and trending north-south. A scarp about 1 m high marks the trace. Tension cracks oriented northeast attest to right-lateral faulting. Khil'ko and others (1985) measured a right-lateral offset of 2.5 to 3 m in a few places and listed an average slip of 2.5 m. They assigned the Bij rupture an age of 500 to 1,000 yr. In addition, Tikhonov (1974, p. 200) reported a significant vertical offset on this fault since it became active in Cenozoic time.

A bound on the rate of right-lateral slip in the Mongolian Altay

Several right-lateral faults trend northwest to north-northwest parallel to the Mongolian Altay and are clear on the Landsat imagery (Fig. 5) (Tapponnier and Molnar, 1979). Several are marked by clear recent fault scarps mapped in detail by Khil'ko and others (1985) and by Shi and others (1984). Here we use the average displacements and lengths of ruptures to estimate the average slip rate across the Mongolian Altay.

The sum of seismic moments of earthquakes along a fault can be used to obtain an estimate of the rate of slip on the fault

(Brune, 1968). The slip rate is the quotient of that sum divided by the shear modulus, the length and the width of the fault, and the time spanned by the earthquake history. We may assume that the faults in the Mongolian Altay are parallel and use the product of the average slip and the length of a rupture as a surrogate for the seismic moment. Hence, we avoid the calculation of seismic moments and the subsequent division of their sum by the shear modulus and the width of the faults. The relevant products are given by: 8 m along 180 km (Fu-yun rupture), 2.5 m along 36 km (Sagsay rupture), and 4.5 m along 215 km (Ar Hötöl rupture). The sum of the products of these ruptures and lengths is 2,500 km · m. Dividing this by an overall length of the Mongolian Altay of 500 km and by 500 yrs, the approximate period in which these ruptures occurred, we obtain an average slip rate of 10 mm/a. This estimate could be an overestimate, if the last 500 years of seismicity were unusually high and unrepresentative of a longer duration, or if the ages of ruptures assigned by Khil'ko and others (1985) were too small. On the other hand, because the region is very remote, it is also likely that evidence for some previous earthquakes has not yet been found or has been destroyed by erosion. We suspect that the rate is not likely to be smaller than 5 mm/a, especially given reported estimates of slip-rates of 10 mm/a for the Fu-yun fault (Shi and others, 1984) and 6 mm/a for the Hovd fault (Trifonov, 1985) and the clear offsets of late Quaternary valleys along the Hovd fault (described above).

Left-lateral faulting at the southern end of the Mongolian Altay

The zone of anastomosing right-lateral faults parallel to the Mongolian Altay also includes conjugate left-lateral faults. A left-lateral fault at the southwest end of the Mongolian Altay, the Bulgan fault (Figs. 3 and 4), has been studied in some detail. Khil'ko and others (1985, p. 31–32) traced a surface rupture due east (090°) for 30 to 32 km from the border with China to about 46.2°N, 91.4°E, noting that the length is longer if it continues far into China. Large mole tracks with heights of 2 to 3 m, and rarely 4 to 5 m, attest to a large left-lateral component. Much of the trace is marked by a south-facing scarp, as high as 1.5 to 2 m (apparently, reverse faulting on a northerly dipping fault), but with an opposite sense in the western part. Khil'ko and others (1985) listed average horizontal and vertical displacements of 2 m and 1 m, respectively, and assigned a tentative age of 500 to 1,000 yr to the surface faulting.

Tikhonov (1974, p. 203–205) discussed the region including the Bulgan rupture, its eastern continuation, and the area to the south in some detail. The fault zone is not simple or narrow. Tikhonov (1974) described evidence for strike-slip and reverse faulting, in which several splays define a wide, east-west–trending zone. The Bulgan fault itselt dips north and is clearly associated with thrust faulting of metamorphic rock onto the Baruun Haraa basin to the south. Tikhonov (1974, p. 205) associated the metamorphism with slip on the fault. He reported overthrusting of several tens of kilometers, but it is not clear what fraction of this slip occurred in Cenozoic time.

Within the Baruun Huuray basin, south of the Bulgan fault, the most significant fault—the Baruun Huuray fault—trends east-northeast and intersects the Bulgan fault east of the town of Bulgan. Its northern continuation is defined by a series of subparallel splays with a cumulative offset of 7 to 8 km (Tikhonov, 1974, p. 205). Along the segment south of the Bulgan fault, Tikhonov reported left-lateral offsets of 15 km of Devonian rock and of 300 to 400 m of middle to late Pleistocene sediment, implying that the fault is currently active with a slip rate of a few mm/a.

It is worth noting that both the Bulgan and the Baruum Huuray faults can be seen on the Landsat imagery, but neither stands out as a major fault (Tapponnier and Molnar, 1979).

The Tahiynshar earthquake, 1974 July 4, in southwestern Mongolia (45.14°N, 94.03°E, M = 7.0) provides a second example of left-lateral slip near the southern extremity of the Mongolian Altay. The fault plane solution for this event shows nearly pure strike-slip faulting: left-lateral on a plane striking 078° or right-lateral on the perpendicular plane (Fig. 2) (Huang and Chen, 1986; Tapponnier and Molnar, 1979). Khil'ko and others (1985, p. 74) reported a zone of tension cracks and compressive features extending in an east-northeast direction for 17 km (Fig. 4). The width of the zone reaches 15 m, and some cracks opened as much as 0.5 m. From the orientations and dimensions of these cracks and mole tracks, they inferred left-lateral slip of 0.3 to 0.4 m. They also reported a maximum vertical slip of 0.4 m but with a small average of 0.1 m. The seismic moment obtained from a fault length of 17 km, width of 20 km, and average slip of 0.3 to 0.4 m is only 3 to 4×10^{18} N m, a small value for an earthquake with a magnitude of 7 and in poor agreement with the value of 8.5×10^{18} N m obtained by Huang and Chen (1986) from syntheses of long-period body waves. If 0.3 to 0.4 m of opening across tension cracks, however, were due to strike-slip faulting, the slip could be 0.6 m. An average of 0.6 (± 0.3) m of left-lateral slip on a plane 20 km long, slightly longer than that measured by Khil'ko and others (1985), yields a scalar seismic moment of $8 (\pm 4) \times 10^{18}$ N m.

Left-lateral faulting is also clear at the southeast end of the Mongolian Altay, where it meets the Gobi Altay. The Shargyn fault (near 46.5°N, 95.5°E) is very clear on the satellite imagery (Figs. 23, 24, and 25). This fault bounds the southern end of the Darviyn Nuruu and forms an arc on the north side of the Shargyn Tsagaan basin. A clear scarp can be seen along the eastern part of the fault for a distance of 55 km. For about 30 km east of Tajgar Bulag to the Hoit Shargyn Gol (Fig. 23), a scarp faces north and crosses young alluvial deposits. In places the height of the scarp reaches 2 to 2.5 m, but along much of this segment its height is only 1 to 1.5 m (Fig. 26). Farther east, along Buuralyn Nuruu (Fig. 25), the fault is marked both by a scarp and by a wide zone (10 to 20 m) of highly fractured, or brecciated, rock. In 1988, two of us (A. B. and M. G. D.) and V. V. Ruzhich measured southward dips of 35 to 55° in bedrock exposures and small (~1 m),

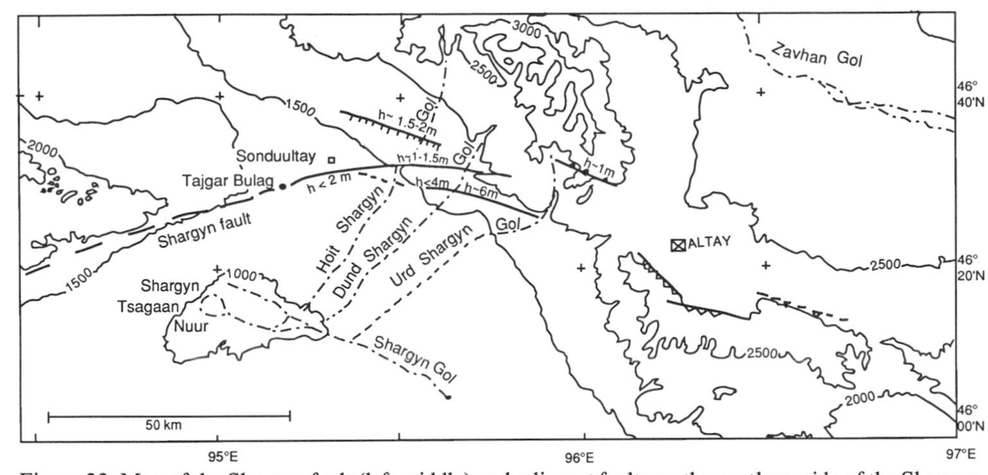

Figure 23. Map of the Shargyn fault (left middle) and adjacent faults on the northern side of the Shargyn Tsagaan basin. Fine lines show elevation contours in meters, and darker lines show surface ruptures. Teeth and hatching on the ruptures point in the dip directions, with teeth showing reverse and hatching normal faulting. All other numbers show heights of scarps. The photo in Figure 26 shows the Shargyn fault at its east end, between the Hoit and Dund Shargyn gols. The photo in Figure 30 shows the reverse scarp northwest of Altay.

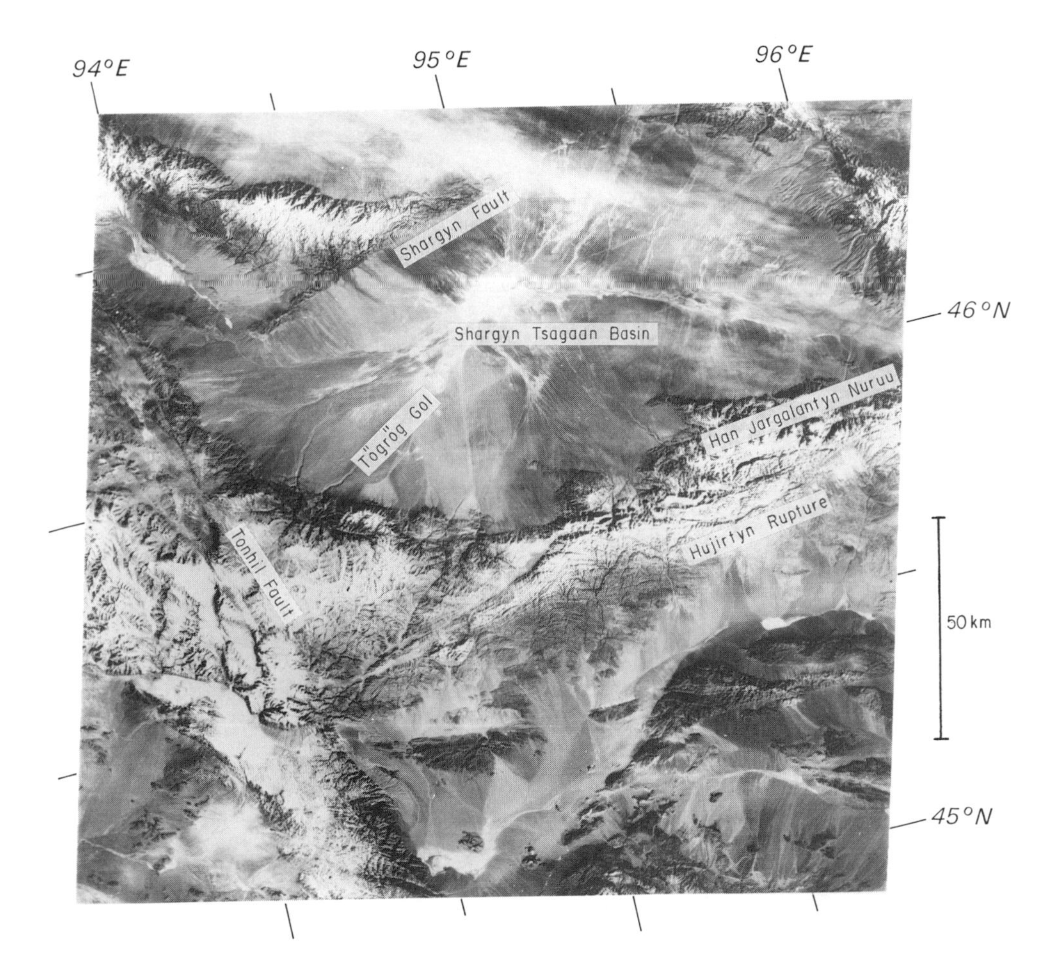

Figure 24. Landsat image 1502-03575-5 of the Shargyn fault (northwest part of the image), the Shargyn Tsagaan basin (north center), and the surrounding area. The Tonhil fault trends north-south across the western part of the image. The Hujirtyn fault casts sharp shadows near 45.75°N, 95.50°E. The Han Jargalantyn Nuruu crosses the middle right part of the image just south of 46°N latitude. The north part of this region is shown on the map in Figure 23. (Modified from Tapponnier and Molnar [1979].)

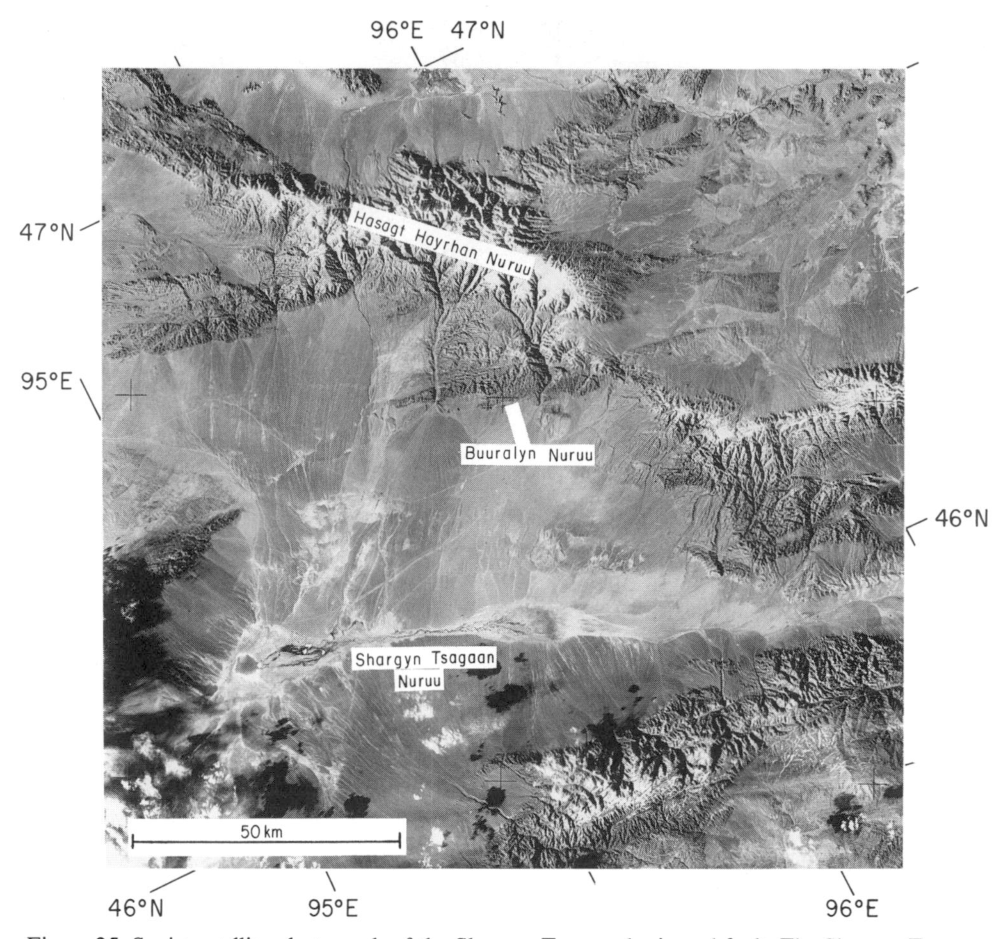

Figure 25. Soviet satellite photograph of the Shargyn Tsagaan basin and fault. The Shargyn Tsagaan basin lies in the lower center of the image, and the Shargyn fault enters from the west near 46°25′N. The Hasagt Hayrhan Nuruu is marked by the snow-capped high terrain in the upper middle part of the image. Note the sharp breaks in slope at the foot of the range, suggestive of active faulting. Some of this region is shown on the map in Figure 23 and the Landsat image in Figure 24.

Figure 26. Photo of the Shargyn rupture near its eastern end. View is southwest, and the scarp faces north here, approximately 1.5 m high, between black arrows. A. B. stands on the scarp and M. G. D. below it. This photo was taken between the Hoit Shargyn Gol and Dund Shargyn Gol (Fig. 23). (Photo by P. Molnar, August 1991.)

but consistent, left-lateral offsets of divides between dry streambeds on the north side of Buuralyn Nuur. Large ridges oriented northwest-southeast between Tajgar Bulag and the village of Sonduultay (Fig. 23) appear to be pressure ridges also indicating left-lateral slip.

A second, roughly parallel active fault lies south of the Buuralyn Nuur and is clear for about 40 km (Figs. 23 and 25). This arcuate scarp faces south, and the causative fault appears to dip north. Near the Dund Shargyn Gol, the height of the scarp reaches 6 m, and farther west it is as high as 4 m. Among us, only M. G. D has visited this area. He considers the scarp to be due to surface faulting but recognizes that the height might be exaggerated by subsequent erosion of its base.

A third scarp is clear northwest of the Buuralyn Nuruu, especially where it crosses the Hoit Shargyn Gol (Figs. 23 and 25). This scarp faces south and is about 1.5 to 2 m high. Basement rock is clearly displaced; in the steep walls of the Hoit Shargyn Gol, 6 to 8 m of Neogene coarse sandstone overlying Precambrian metamorphic rock south of the fault are in fault contact with the Precambrian rock to the north. The fault dips steeply, but the value cannot be measured reliably. The scarp is not straight, suggesting either that it is a fault-line scarp or that the strike-slip component is minor.

All of these surface ruptures are clear and in many places sharp. The absence of free faces on the scarps and the gentle dips of most of them, however, imply that the ruptures are relatively old. We suspect that most are older than 1,000 yr.

Reverse or thrust faulting in the Mongolian Altay

As noted above, examples of reverse faulting are fewer and less impressive than those of strike-slip, particularly right-lateral, faulting. Reverse faulting plays a major role in creating the large-scale topography, whereas surface ruptures rarely are impressive, perhaps because many are blind or are covered quickly by sediment from the adjacent highlands.

A short west-northwesterly trending fault zone, the Tsagaan Shuvuut fault zone, is clear on the Landsat imagery of the northern end of the Mongolian Altay (Tapponnier and Molnar, 1979). Devyatkin (1974, p. 191; 1975, p. 267) reported southwestward overthrusting of Paleozoic rock onto Jurassic sedimentary rock and of both onto Neogene deposits. This zone is marked by an abrupt rise in the mountain front and by an eroded scarp that can be discerned on some of the fans formed where streams debouch from the mountains (Fig. 27).

Faulting within this zone is likely to have occurred in the Üüreg Nuur earthquake, of 1970 May 15 (50.17°N, 91.25°E, M = 7.0) (Figs. 2 and 4). The fault plane solution for this earthquake is poorly constrained by P- and S-wave first motions but shows a large component of reverse, or thrust, faulting (Tapponnier and Molnar, 1979; Vilkas, 1982). Khil'ko and others (1985, p. 72–74) reported a zone of cracks for a distance of 6 to 8 km trending roughly east-west. The north side rose with respect to the south side as much as 2 m (Fig. 28), but the segment with impressive surface deformation is very short, less than 1 km in length. Along much of the zone the deformation consists of tension cracks (Fig. 29), in some cases with openings of up to 4 m. Khil'ko and others (1985) also reported an "insignificant left-lateral component." Although left-lateral slip in some areas seems to be large (Fig. 28c), such areas are localized and not necessarily typical of the entire rupture zone.

This deformation does not appear to represent primary faulting. Instead it resembles the complex deformation in the hanging walls of the ruptures associated with the 1981 El Asnam earthquake in Algeria (Avouac and others, 1993; King and Vita-Finzi, 1981; Philip and Meghraoui, 1983) and the 1988 Spitak earthquake in Armenia (Philip and others, 1992). We suspect that it is not representative of the faulting that occurred at depth.

From a synthesis of P waveforms, Vilkas (1982) inferred largely reverse faulting but with a substantial left-lateral strike-slip component on the northerly dipping nodal plane. Her seismic moment of 1×10^{19} N m is consistent with an average of 1.5 m of oblique reverse faulting occurring on a plane dipping 58° north and with an east-west dimension of 10 km (1.2×10^{19} N m).

Figure 27. Photographs (a: distant view; b and c: close-ups with 135-mm lens) of fault scarps at the base of the Shuvuut Tsagaan range north of Üüreg Nuur (Fig. 4). Views in a and b are north-northwest; that in c is west-northwest. Note the dissection of perched alluvial fans above the valley floor, at the base of the range. Apparently, slip on a reverse fault dipping north beneath the range has elevated the fans. The photo in c shows an area west of that in a and b; Üüreg Nuur lies in the foreground of c. (Photos by P. Molnar, August 1991.)

The reverse component of slip and the nearly east-west orientation of this fault zone suggest that the right-lateral slip along the north-northwesterly trending faults in the Mongolian

Figure 29. Surface deformation associated with the 1970 Üüreg Nuur earthquake. View is west down a steep hill immediately west of the area in Figure 28c and shows a large component of opening perpendicular to the orientation of the fault. This deformation can be traced to the bottom of the valley in the upper middle of the photo but not across to the other side. (Photo by P. Molnar, August 1991.)

Figure 28. Photos of the surface rupture of the 1970 Üüreg Nuur earthquake, where surface deformation was maximum. Views are north-northwest in a, west in b, and north-northeast in c. Notice that the minor slump on the scarp in the left of a is in the middle of c. The vertical component is associated with a component of extension, as can be seen in b. The apparent left-lateral displacement in c is not obvious as close as 100 m in either direction and is due, in part, to the large (1 to 2 m) component of opening across the rupture (see also Fig. 29). (A. B. in a and b and M. G. D. in c provide scales.) (Photos by P. Molnar, August 1991.)

Altay is absorbed, at least in part, by crustal shortening at the northwest terminations of these faults. Such deformation is consistent with the Mongolian Altay deforming largely by right-lateral strike slip such that the Mongolian Altay rotates counterclockwise. Faults like the Tsagaan Shuvuut accommodate the terminations of the strike-slip faults required by rotations of blocks.

Thrust faulting with northeast-southwest shortening also occurs at the southeast end of the Mongolian Altay. In 1991, three of us (A. B., M. G. D., and P. M.) traced a clear scarp for about 15 km along the foot of the Hasagt Hayrhan Nuruu, north of the Shargyn Tsagaan basin (Figs. 23 and 25). The scarp faces south and is roughly 1 to 1.5 m high (Fig. 30). Deep river valleys crossing the edge of the range and the fault expose a relatively steep (40 to 60°) northeastward dip, indicating a component of reverse faulting. Farther northwest, the sharp break in slope along the southern edge of the Hasagt Hayrhan Nuruu, clear on the satellite photo (Fig. 25), suggests active thrust faulting.

In 1988, one of us (M.G.D.) studied another young thrust fault (the Tögrög fault) near the southern end of the Mongolian Altay. Near 46°N, 94.75°E an intermittent scarp, up to 6 m high,

Figure 30. Photograph of a reverse fault scarp just south of the Hasagt Hayrhan Nuruu and northwest of the town of Altay (see Figs. 23 and 25). View is north, and the scarp (between open arrows) is about 1 to 1.5 m in height. Jeep on left and people in the middle (beneath black arrows) provide a scale. (Photo by P. Molnar, August 1991.)

can be traced southeastward for about 40 km. It lies about 10 to 15 km northeast of the very abrupt front of the Mongolian Altay where it bounds the Shargyn Tsagaan basin (Fig. 24). Even where young gullies have dissected the scarp it stands 1 to 1.5 m high. An ephemeral stream, the Tögrög Gol (Fig. 24), has cut a sharply defined canyon some 15 m deep into the Tertiary sedimentary rock and overlying Quaternary deposits, exposing a cross section in the walls of the canyon. Young sand and gravel near the earth's surface dip south at 10 to 12°, despite the gentle northerly dip of the surface. Thus, these young deposits have been tilted. They overlie cemented gray conglomerate and gravel, typical of Paleogene sedimentary rock, that in turn overlie in fault contact red siltstone and clay, presumably of Neogene age. In addition, gentle warping of a river terrace, 4 m above the floor of the valley, is apparent. The warp is as much as 3 m high and is spread over an area about 75 m wide. The valley floor is warped up only 0.5 to 0.75 m, suggesting that the warping is active. Both the localized warping and the southward dip of the Quaternary deposits suggest that the fault is listric and flattens to the south. This thrust fault is probably only a splay from the main thrust fault that bounds the range 10 to 15 km to the southwest. Nevertheless, the relatively high scarp, as high as 6 m in places, and the gentle dip of the fault imply that thrust slip and convergence are not slow.

Probably with further work, other examples of active thrust faulting and northeast-southwest crustal shortening along the margins of the Mongolian Altay will be found, but at present the evidence for strike-slip faulting is much more obvious.

Summary

The main style of faulting in the Mongolian Altay appears to be right-lateral strike slip on planes trending north-northwest. As noted above, from surface ruptures alone, the rate of right-lateral slip across the range appears to be at least a few mm/a. Given the prominence of the faults, the possibility that some surface ruptures have not yet been found, and the clear offsets of a few kilometers on them, the rate could exceed 10 mm/a. In addition, however, some crustal shortening also occurs across the range, as indicated by fault plane solutions of some earthquakes (Fig. 2) and sparse field observations, manifesting itself as a high range of mountains. The limited seismic history and the few reports of Holocene faulting, however, do not allow the rate of this process to be quantified. If the rate of strike-slip faulting is 10 mm/a, shortening at a rate between about 2 and several mm/a seems likely. At the northwest end of the Mongolian Altay, the main northwest-trending strike-slip faults seem to terminate in thrust and reverse faults. At its southeast end, these strike-slip faults appear to die out or terminate against other faults, and left-lateral slip on the conjugate, easterly trending planes becomes the dominant mode of deformation. This style prevails in the easterly trending Gobi Altay range that continues to the east-southeast.

THE GOBI ALTAY

As noted above, the Gobi Altay might be seen as the southeastward geographic continuation of the Mongolian Altay, but we isolate it here because the dominant sense of active deformation is so different from that of the Mongolian Altay farther west. Our knowledge of the deformation in the Gobi Altay is provided largely by the thorough investigation by V. P. Solonenko and his colleagues (Florensov and Solonenko, 1963) of the surface deformation associated with the 1957 Gobi Altay earthquake and the Bogd fault. Let us begin with a discussion of that earthquake and then consider the surrounding regions.

The Gobi Altay earthquake of 1957 December 4 (45.31°N, 99.21°E, M = 8.3)

The comprehensive investigations by Solonenko and his colleagues, immediately following this event while the ground was still frozen (Solonenko and others, 1960) and again in the following autumn (Florensov and Solonenko, 1963), make this one of the world's most thoroughly studied surface ruptures of a great earthquake. Very large components of reverse and left-lateral strike-slip faulting characterize a zone roughly 250 km long and more than 30 km wide (Fig. 31). Moreover, the rupture zone is not straight and includes numerous jogs and splays. The maximum reported strike-slip and vertical components approach 7 to 8 m and 9 m, respectively, but the components of displacement vary markedly along the rupture.

Three main zones of surface faults comprise most of the surface faulting (Fig. 31). The main surface rupture, along the Bogd fault, trends roughly east-southeast for 250 km along the northern flank of the mountain Ih Bogd and its continuations east and west. A second rupture, the Toromhon Overthrust, trends roughly north-south for 32 km (Solonenko and others, 1960) and approaches the Bogd fault from the south in the eastern third of the rupture. The third break lies along the southern flank of Ih Bogd and trends roughly west-northwest for a total length of about 70 km.

Bogd fault. The average strike is about 100°, but the surface trace is by no means straight. The trends of individual segments span the range from northeast to southeast. As a general rule, but with many exceptions, those segments with strikes close to 090° show nearly pure strike-slip displacements, with vertical components smaller than 1 m (Fig. 32). Vertical components are largest along segments striking east-southeast (095 to 100°) or southeast. In such cases, the southern side is consistently the upthrown block (Figs. 33, 34, and 35). The north side forms the upthrown block in a few areas (Figs. 32a and 36), where the strike of the rupture is slightly east-northeast (080°). This pattern suggests that the horizontal component of the slip vector trends nearly east-west.

Virtually everywhere that it was possible to measure or to estimate the dip of the Bogd fault, Florensov and Soloneko (1963) measured a southward dip. Both L. M. Balakina's fault

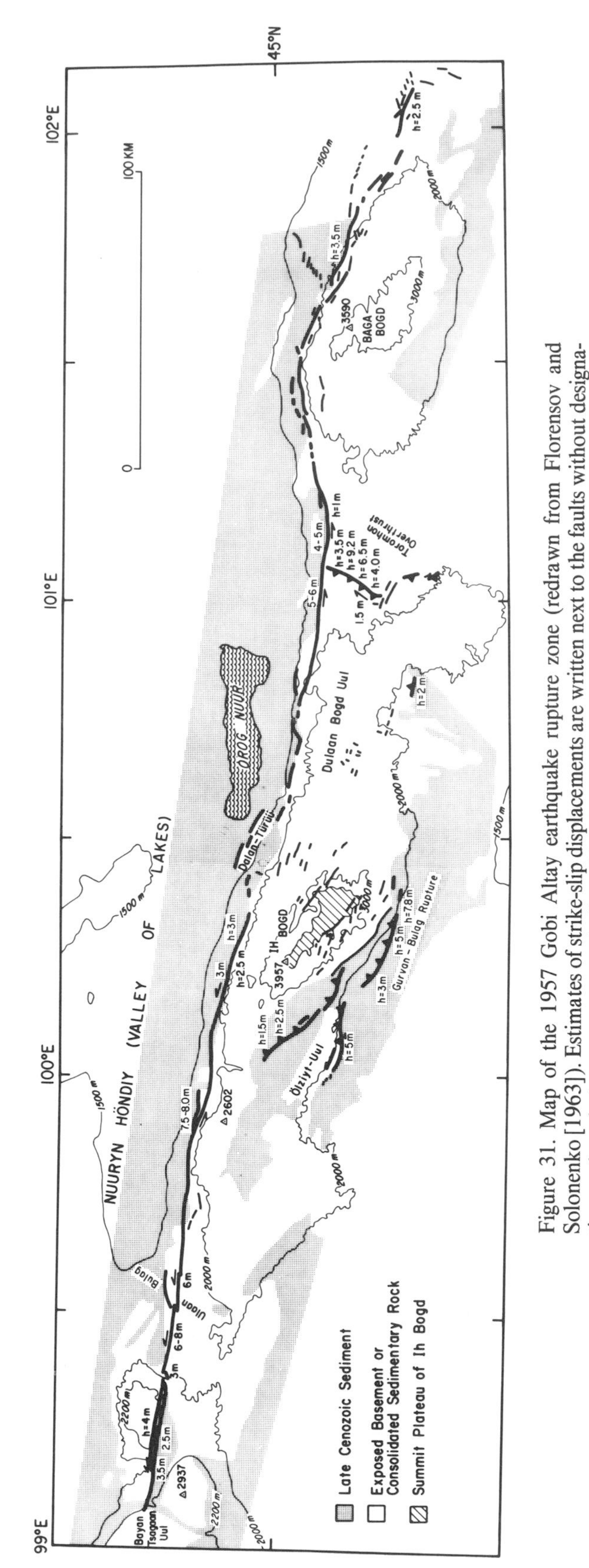

Figure 31. Map of the 1957 Gobi Altay earthquake rupture zone (redrawn from Florensov and Solonenko [1963]). Estimates of strike-slip displacements are written next to the faults without designation, and vertical components are indicated by the letter h.

Figure 32. Photographs of Gobi Altay earthquake rupture a few kilometers east of Ulaan Bulag (Fig. 31). Views are east-northeast in a and southwest in b. Left-lateral displacements in this area reach 6 m. Vertical displacements rarely exceed 1.5 m. Notice that in a the north side has been uplifted, but in b and in Figure 33 it has been downdropped relative to the south side. (Photo a by P. Molnar and photo b by H. Philip, July 1990.)

Figure 33. Photograph of Gobi Altay earthquake rupture a few kilometers east of Ulaan Bulag (Fig. 31) and about 100 m west of the view in Figure 32b, showing a vertical component (between arrows) of about 1.2 m. View is southeast. H. P. provides the scale. (Photo by P. Molnar, July 1990.)

Figure 34. Photographs of Gobi Altay earthquake rupture near the area of Solonenko's maximum slip, near where the Toromhon Overthrust approaches the Bogd fault (Fig. 31). Views are downhill toward the south in a and southwest in b. We infer a horizontal component of approximately 4 m and a vertical component of about 1.5 m, with the south side up. Figures 40 and 41 show different views of the two gullies in b. (Photos by P. Molnar, July 1990.)

plane solution (reported by Florensov and Solonenko [1963]) and Okal's (1976) improvement of it show a southerly dip for the initial rupture. Florensov and Solonenko (1963) stated that where they could measure it (at the surface, of course), the dip is consistently between 65 and 70°, which is somewhat steeper than the values of 30 to 55° inferred by Balakina and 53° by Okal (1976). If a southerly dip does apply to all segments of

Figure 35. Photograph of Gobi Altay earthquake rupture approximately 1 km west of where the Toromhon Overthrust approaches the Bogd fault (Fig. 31). View is east-northeast and shows a vertical component in alluvium of about 1.2 m. R. A. K. provides the scale. The Toromhon Sayr (shown also in Fig. 48) is the streambed in the foreground. (Photo by P. Molnar, July 1990.)

Figure 36. Photograph of Gobi Altay earthquake rupture west of where the Toromhon Overthrust approaches the Bogd fault (Fig. 31). View is north at a low south-facing scarp, approximately 0.4 m high, between black arrows. A. C., on the left, and H. P., in the center, mark the truncations of a shallow gully offset horizontally about 5 m. This gully enters the photo from the lower left corner and recedes behind A. C. toward the center of the photo. R. A. K. and A. B., on the right, stand on the northern, upper block, and L. G. farthest right, stands in front of it. (Photo by P. Molnar, July 1990.)

the Bogd surface rupture and if the horizontal component of the slip vector is oriented approximately east-west, most vertical components of slip mark oblique reverse faults, and the few cases where the north side is upthrown indicate oblique normal faulting. In estimating a seismic moment, we assumed a uniform southerly dip of 55°.

The magnitudes of strike-slip and vertical components are qualitatively consistent with a roughly east-west slip vector, insofar as segments of the fault consistently dip south. Consider the oblique slip shown in Figure 37; Δu_h is the horizontal component of slip on the fault. Where slip is purely strike-slip, the strike-slip component, Δu_{ss}, should equal Δu_h. Where the strike of the fault deviates from the orientation of the horizontal component of the slip vector, $\Delta u_{ss} = \Delta u_h \cdot \cos \Delta\theta$, where $\Delta\theta$ is the angle between the azimuth of the slip vector and the strike of the segment of surface rupture. For small values of $\Delta\theta$, $< 20°$, $\Delta u_{ss} \cong \Delta u_h$. The vertical component, Δu_v, is related to the horizontal convergent component perpendicular to the fault, Δu_c, by $\Delta u_v = \Delta u_c \tan \delta$, where δ is the dip of the fault (Fig. 37). For instance, for a dip of 63°, $\Delta u_v = 2\ \Delta u_c$. Finally, $\Delta u_c = \Delta u_{ss} \cdot \tan \Delta\theta$. Hence the vertical and strike-slip components should be related by $\Delta u_v = \Delta u_{ss} \cdot \tan \delta \cdot \tan \Delta\theta$ (Fig. 37). Suppose that the horizontal component of the slip vector pointed in the direction 085° and that the horizontal slip were 6 m. For strikes of 085°, there would be no component of slip perpendicular to the fault and no vertical component. For strikes of 095° and 102°, and for a dip $\delta = 55°$, vertical components should be 1.4 and 2.5 m, respectively. Similarly, for segments striking 080°, there should be a vertical component of normal faulting of about 0.75 m. We note these values because they are close to what we observed in the eastern third of the rupture.

Let us discuss variations in the character of the Bogd fault from west to east. Solonenko and others (1960) assigned the westernmost rupture to 45°16′N, 98°52′E and showed discontinuous ruptures on their map in this region, but later work apparently did not corroborate these features, seen only from the air in January 1958. The westernmost rupture shown by Florensov and Solonenko (1963) is a few kilometers east of the 99°E meridian. In this westernmost segment, separate splays surround and obliquely cross a narrow graben, 350 to 800 m wide. Solonenko and others (1960, p. 29–35) reported a general northeasterly orientation of individual ruptures and ascribed much of the displacement to normal faulting. The bounding scarps were commonly as high as 2 to 2.5 m, but those within the graben were typically only 0.2 to 0.5 m in height. Subsequent erosion has virtually obliterated evidence of internal deformation of the floor of the graben, but the bounding scarps are clear in many localities. It is difficult to measure the total strike-slip or vertical component.

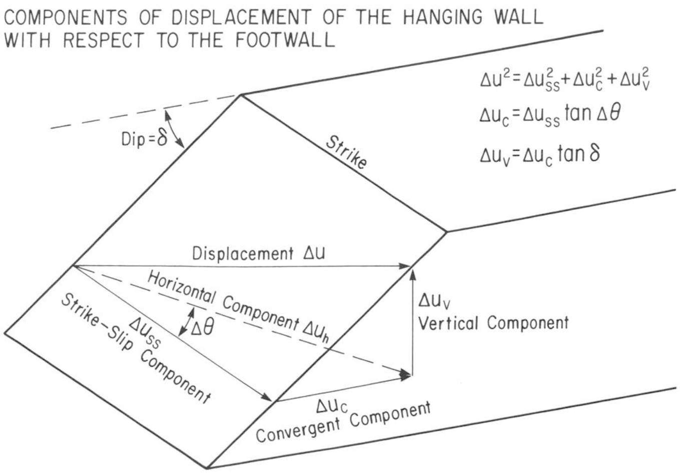

Figure 37. Block diagram showing the relationships of horizontal and vertical components of displacement to the slip vector and local strike of the fault.

Farther east, just east and west of Ulaan Bulag (Fig. 31), however, the surface rupture is localized on one fault. A very spectacular surface rupture with about 6 m of strike-slip displacement and only a small vertical component (Fig. 32) trends roughly 090 to 095°. In a few localities, the rupture steps roughly 100 m in either right-stepping or left-stepping configurations. In the overlapping zones, particularly where the strike is somewhat north of east, small grabens have formed (Fig. 38). Small differences in the strike are associated with differences in vertical components of slip and with the directions that the scarps face. Where the strike is between 080 and 100°, the scarp faces south. Where the fault strikes about 105° or more southeasterly, the scarp faces north and commonly is 1 to 1.5 m high (Fig. 33).

Florensov and Solonenko (1963) also reported an offset of 6 m for this area, but Trifonov (1985, 1988), for reasons that are not clear to us, associated only smaller offsets with the earthquake in 1957. He reported only 2.7 to 3.3 m for localities near those in Figure 32 (Trifonov, 1985, p. 19). Later he noted a "massive number of measurements" with a maximum of about 3 to 3.5 m for Ulaan Bulag area (Trifonov, 1988, p. 257). Trifonov (1988, p. 256) stated that the "total strike slip associated with the earthquake of 1957 nowhere exceeds 5 m." We disagree strongly with him.

Near 45.1°N, 101°E, the overall trend of the Bogd fault becomes more southeasterly to about 105°, and it is along this segment that both horizontal and vertical components appear to be the largest. Florensov and Solonenko (1963) reported horizontal components as large as 7 to 8 m and vertical components of 2 to 3 m. As in the west, the rupture is not straight, and in places more than one strand ruptured. In one locality, just west of the lake Orog Nuur, the rupture steps to the north and seems to bound a young, growing linear hill, Dalan Türüü (Fig. 31), where Plio-Quaternary conglomerate, breccia, and sandstone crop out.

Farther east, just north of the Dulaan Bogd Uul, the rupture trends more easterly (095°), the strike-slip component is 5 to 6 m (Fig. 39), and the vertical component is only about 1 m. As in the area northwest of Ih Bogd, slight variations in the local trend are associated with variations in the vertical component of slip.

Where the local strike is greater than roughly 090°, the scarp faces north, commonly with a height of 1 m or more (Figs. 33 and 34). Where the strike is less than 085 to 090°, a south-facing scarp exists (Fig. 36), with a height of less than 1 m. Thus the slip vector here may be oriented more to the north of east than it is farther west.

The amplitude of the strike-slip component appears to decrease (from 5 to 6 m to about 3 to 4 m) east of where the Toromhon Overthrust approaches the main rupture (Florensov and Solonenko, 1963). The main rupture crosses the south flank of a low ridge and separates resistant Paleozoic granite on the north from less-resistant Cretaceous red sandstone on the south. In the segment about 1 km long where the Toromhon Overthrust approaches the Bogd fault, the amount that small gullies are offset varies from one dry gully to the next. Numerous dry valleys are displaced left-laterally about 4 to 5 m, and the southern side has been uplifted about 1 m with respect to the northern side (Fig. 34). The western margins of two of the largest, southerly flowing, dry gullies have been displaced 8.85 and 8.6 m left-laterally, however, and Solonenko seems to have associated these large offsets with the Gobi Altay earthquake of 1957 (Fig. 40). These are the largest offsets that Florensov and Solonenko (1963) reported along the Bogd fault. We do not dispute the observed offsets, but we observed that the large displacements are not apparent on the eastern flanks of the same dry streambeds. Moreover, the smaller offsets of 4 to 5 m of adjacent smaller valleys make it difficult to explain a difference of 4 to 5 m in offset from one gully to the next (Fig. 41). Thus, we, like Trifonov (1988), suspect that only about half of these 8- to 9-m offsets occurred in 1957 and that these large values record two earthquakes. Because these offsets are associated with the widest and deepest of the dry

Figure 38. Photographs of Gobi Altay earthquake rupture approximately 1 to 2 km west of Ulaan Bulag (Fig. 31). a and b show two parallel ruptures and a connecting fault in a right-stepping relay. a, View is north across the region in the background of b. The southern trace in the foreground of a dies out to the west. A connecting fault, between Yu. Ya. V. in the foreground and R. A. K. on the left and denoted by diagonal black arrows, transfers slip to the northern trace, indicated by dark downward-pointing arrows in upper right. Farther west, the northern trace becomes dominant. b, View east shows the two traces in a in the distance (black arrows). The southern (right) trace continues farther to the east but dies out to the west (right). The northern trace becomes dominant in the foreground. c, Photo looking north and taken about 100 m west of the area in b. Here, only a single trace can be recognized. The measured left-lateral offset is 7 ± 2 m. B. A. B. stands on the northern segment of the displaced ridge. (Photos by P. Molnar, with lenses of 50 mm [a], 135 mm [b], and 20 mm [c], July 1990.)

gullies displaced by the Bogd fault in this area, they are the gullies most likely to preserve evidence of more than one rupture.

Farther east, the rupture wraps around the northern flank of the Baga Bogd range. On the northwest flank, the strikes of surface ruptures lie between 075 and 090°, and vertical components are small (⩽ 1 m). On the northeast flank, strikes are between 090 and 135°, and Florensov and Solonenko (1963) reported vertical components of 2 to 3.5 m. Although they indicated left-lateral strike-slip displacements, they gave no values. The variation of the vertical components with the local strike is consistent with an easterly slip vector. Solonenko and others (1960) placed the east end of the rupture at 44°48′N, 102°10′E, making the length about 250 km.

Figure 41. Photograph of the gully adjacent to and about 30 m west of the one shown in Figure 40 (see also Fig. 34b). View is south and downhill. R. A. K. (left) and A. C. mark the thalwegs of the gully, south and north of the fault, respectively. The offset is only about 4 to 5 m, so much less than that in Figure 40 that we suspect that the offset shown in that figure occurred by two earthquakes. (Photo by P. Molnar, July 1990.)

Figure 39. Oblique aerial photo of Gobi Altay earthquake rupture looking east-southeast from south of Orog Nuur (Fig. 31). (From Florensov and Solonenko [1963, Fig. 113]).

Figure 40. Photograph of an offset gully, near where the Toromhon Overthrust approaches the Bogd fault (Fig. 31). View is south and downhill. Solonenko, in Florensov and Solonenko (1963), attributed this entire 8.85-m horizontal offset, the maximum that he observed, to the Gobi Altay earthquake of 1957. We, however, suspect that this offset occurred in two steps (see Figs. 34b and 41). R. A. K. and A. C. mark the upstream and downstream positions of the west sides of the gully, respectively. (Photo by P. Molnar, July 1990.)

Toromhon Overthrust. This segment approaches the main rupture from the south just east of the 101°E meridian. The overall trend of the scarp is roughly north-south, but considerable variability exists, including one segment that trends roughly east-southeast (Fig. 31). The scarp consistently faces east. At the surface, it appears to dip steeply west along its northern segment, but farther south, gentler dips seem to be required by its trace across the topographic contours. Measured heights are nearly everywhere greater than 1 m, and Florensov and Solonenko (1963) reported values as large as 6.5 and 9.2 m. Where the scarp traverses relatively level ground, the height is commonly about 2.5 m (Fig. 42). Exceptionally great heights can be observed only where the scarp follows steep easterly sloping topography. Thus, perhaps these large values should be ascribed to previous events or other processes, such as slumping, that also have shaped the topography.

Where the local trend of the scarp is roughly north-south, we saw no clear evidence for strike-slip components, but on northeast-trending segments a right-lateral component is clear. Solonenko and others (1960, p. 39) reported slickensides plunging 58° southwest on a nearly vertical plane striking 030°. This suggests a strike-slip component roughly half that of the vertical component. Florensov and Solonenko (1963) also reported a right-lateral component of 1.5 m where the local strike is about 025°, roughly 15 km southwest of the Bogd fault. Similarly, at the northern end of the overthrust, where the strike of the surface rupture also is 025°, several observations indicate a right-lateral strike-slip component. Some small ridges and dry gullies intersected by the fault show right-lateral separations, with one apparent strike-slip offset of 4 m (Fig. 43). In addition, slickensides at two localities within the fault zone plunge southwest at about 20°, suggesting a greater strike-slip than vertical component. We

Figure 42. Photographs of the Toromhon Overthrust scarp near its northern end (Fig. 31), where the fault strikes nearly due north. Views are to the southwest in a and west in b and show a large vertical component, approximately 3 to 4 m. (Photos by P. Molnar, July 1990.)

Figure 43. Photograph of the Toromhon Overthrust scarp (Fig. 31), where the fault strikes north-northeast, roughly 1 km north of the view in Figure 42. View is to the west-northwest and shows large apparent horizontal (4 m) and vertical (1 m) components. A. C. provides a scale. (Photo by H. Philip, July 1990.)

could not agree among us on the relative importances of strike-slip (between 1 and 4 m) and vertical (1 to 2 m) components along this northernmost segment, but the presence of both components is inescapable. Finally, an (apparently thrust) contact between Paleozoic and Mesozoic rock is offset in a right-lateral sense by about 200 m.

Tapponnier and Molnar (1979) inferred that the Toromhon Overthrust resulted from normal faulting. That inference was based on the apparent juxtaposition of left-lateral offsets along the Bogd fault shown on Florensov and Solonenko's (1963) summary map. That map lists strike-slip displacements of 3.8 m west of where the Toromhon Overthrust approaches the Bogd rupture and 8.85 and 8.6 m east of it. Such a difference in left-lateral slip would require east-west horizontal extension, not compression, across the Toromhon Overthrust. Above, we expressed doubts about ascribing all of the 8.85-m and 8.6-m offsets to the 1957 earthquake. We also suggest that the displacement along the Bogd fault is less to the east of this area (3 to 4 m) than to the west of it (5 to 6 m). Hence, it is consistent with reverse, instead of normal, faulting on the Toromhom Overthrust.

Ruptures south and southwest of Ih Bogd: The Gurvan Bulag zone. Solonenko and others (1960) and Florensov and Solonenko (1963) described a third extensive zone of ruptures that follows the southern and southwestern margins of Ih Bogd and that splays to the west. Solonenko and others (1960) mentioned four separate ruptures with a combined length of 106 km, but later Florensov and Solonenko (1963) noted only tension cracks along one of the rupture zones about 35 km in length. Thus, the cumulative length of this zone of clear surface faulting is roughly 70 km. This length also does not include the extensive deformation of the high terrain of Ih Bogd itself, where the disruption of the surface seems to have been superficial (Florensov and Solonenko, 1963).

The Gurvan Bulag ruptures are marked by high, south-facing scarps. Florensov and Solonenko reported no clear evidence of strike slip along them. The scarps do not lie along the break in slope between the crystalline rock of Ih Bogd and the large alluvial fans surrounding the mountain, as probably would be the case if normal faulting occurred. Instead, the ruptures cut the fans some 1 to 5 km from the break in slope and follow contours in such a way that reverse or thrust faulting must be inferred. Rapid erosion in the summer of 1958 exposed one section through the scarp. V. P. and M. A. Solonenko measured northward dips of 40° near the surface and curving to 30° at a depth of about 8 to 9 m, where the fault offsets "?Oligocene" sedimentary rock dipping 35° to the north (Florensov and Solonenko, 1963, Fig. 130). It probably is unwise to assume that the 30° dip continues to great depth, but a nearly vertical dip also seems unlikely. They also reported a subsurface dip of 68° for the segment just south of the Ölziyt Uul (Fig. 31) but gave no evidence.

The average vertical component of displacement is not well constrained. In a valley exhumed in 1958, V. P. and M. A. Solonenko measured a vertical displacement of 4 to 5 m of the consolidated Tertiary sedimentary rock beneath the unconsolidated fan debris (Florensov and Solonenko, 1963, Fig. 130). Florensov and Solonenko (1963) reported a height of 5 m of the scarp in a couple of other places and 7 to 8 m at one locality.

Because the height of a scarp provides an upper bound on the vertical component of displacement (e.g., Hanks and Andrews, 1989), the vertical component may have been less than these large heights. Where a scarp faces downhill, slumping and degradation of the scarp increase its height (Fig. 44). In 1990, slopes of about 30°, nearly the angle of repose, characterized most scarps in unconsolidated material. Along such reverse faults, all semblance of an overhanging face had been lost. Probably slumping of it occurred shortly after the earthquake. The evolution of a scarp, from initially overhanging to one sloping in the same direction as the surrounding topography, makes the height increase. Slumping of material down from above the initial scarp makes the top go up, and the scattering of this material at the base lowers the bottom of the scarp (Fig. 44). Hence, scarps cut into steeply sloping ground will evolve into higher scarps than those cut into flat surfaces.

We measured the heights of scarps along the Gurvan Bulag rupture that were only about 200 m apart but cut into surfaces with different slopes. Where we measured the slope above and below to be $\alpha = 15°$, the height was h = 6.8 m. Assuming that the dips are $\phi = 30$ to $35°$ for the scarp and $\delta = 49°$ for the fault, the calculated vertical component of displacement is 3.0 to 3.5 m (Fig. 45). Where the surface above the scarp slopes at 25° and below it at 16° (Fig. 45), we measured a height of 8.5 m. For an

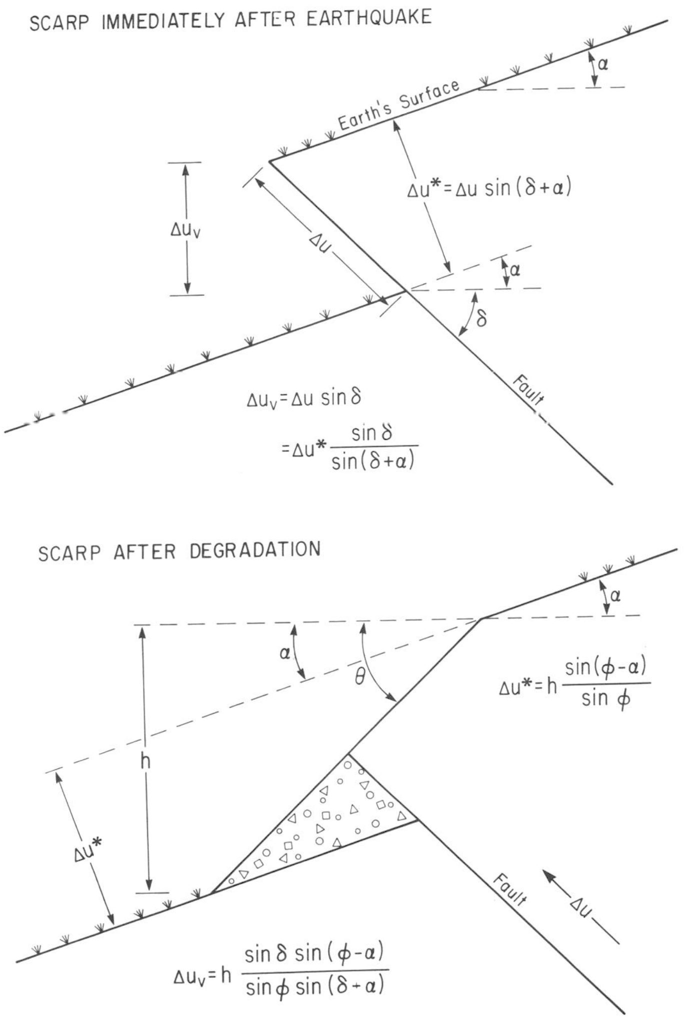

SCARP IMMEDIATELY AFTER EARTHQUAKE

Earth's Surface

$\Delta u^* = \Delta u \sin(\delta + \alpha)$

$\Delta u_V = \Delta u \sin\delta = \Delta u^* \frac{\sin\delta}{\sin(\delta+\alpha)}$

Fault

SCARP AFTER DEGRADATION

$\Delta u^* = h \frac{\sin(\phi-\alpha)}{\sin\phi}$

$\Delta u_V = h \frac{\sin\delta \sin(\phi-\alpha)}{\sin\phi \sin(\delta+\alpha)}$

Figure 44. Simple diagrams showing the degradation of a scarp facing downhill. Upper diagram shows the topography before slumping has occurred. Lower diagram shows how slumping of the hanging wall makes it rise, and the accumulation of material on the footwall makes the bottom of the scarp go down. Thus, the height (h) of the scarp increases and becomes higher than the vertical offset (Δu_V). From the equations shown, the height of the scarp and the slope of the land surface can be used to infer the displacement on the fault.

Figure 45. Photographs of the Gurvan Bulag scarp, which ruptured in the 1957 Gobi Altay earthquake. This scarp (between arrows) lies on the south side of Ih Bogd. Views are northeast. The scarp crosses a huge alluvial fan that extends far to the west of the foreground in a. Measured heights of the scarp are 6.8 m at the left black arrows in a and 8.5 m at the right black arrows in a and in b. The hill above the scarp in a and forming the skyline in b is an especially dissected part of an older fan. It stands about 45 m above its downhill continuation below the scarp. (Photos by P. Molnar, July 1990.)

assumed dip of the fault of $\delta = 49°$ and an average regional slope of $\alpha = 20°$, the estimated vertical displacement for this scarp is 2.5 to 3.0 m. Along much of this segment of surface faulting, 3 (± 1) m seems to be a good estimate for the average vertical displacement.

Seismic moment of the 1957 Gobi Altay earthquake. Let us use the measured displacements to estimate the seismic moment associated with the three ruptures, assuming that they all extended to a depth of 20 km. For simplicity let us assume that slip vector is uniform along the Bogd fault, 085°, and that the average dip is 55° to the south. For average strike-slip and dip-slip components of 6 (± 2) m and 1.4 m in the western segment (65 km long), 6 (± 2) m and 2.5 m in the central segment (100 km long), and 4 (± 2) m and 1 m in the eastern segment (85 km long), M_0 is approximately 1.14 (± 0.4) × 10^{21} N m. (This is approximate because of the different strikes.) If, for the Toromhon Overthrust we assume an average vertical component of slip of about 2.5 m and an average dip at depth of 40°, the average slip is nearly 4 m. With a length of 32 km, the corresponding seismic moment associated with this segment is 0.13 × 10^{21} N m. For the segment south of Ih Bogd, an average vertical component of 3 (± 1) m on a plane dipping 49° corresponds to an average displacement of 4 m. For a length of 70 km, the seismic moment associated with this segment is 0.24 (± 0.06) × 10^{21} N m.

Because of the different orientations and senses of slip on the different segments we cannot add the three scalar values. Nevertheless, it is clear that their combination is within the range of estimates derived from amplitudes of very long period surface waves: 1.4 (± 0.7) × 10^{21} N m (Chen and Molnar, 1977) and 1.8 (± 0.4) × 10^{21} N m (Okal, 1976). Note that smaller dips of the faults, with the same measured strike-slip and vertical components, would yield larger scalar moments. Moreover, we have not included the additional evidence of surface deformation on and near Ih Bogd that might be associated with subsurface faulting.

Estimates of average recurrence intervals of great earthquakes and rates of slip. In two areas we saw evidence of displacements that must have accumulated over thousands of years. Assuming that these displacements occurred since the last glacial maximum (20 ka) (Bard and others, 1990), since the beginning of the Holocene epoch (10 ka), or since 5 ka, they can be used to estimate ranges of average rates of slip and average recurrence intervals for earthquakes comparable with that in 1957.

In the segment of the Bogd fault east of Ulaan Bulag, where the surface rupture is very sharply defined (Figs. 32 and 33), recent incision by a northerly flowing stream has exposed a vertical offset of both unconsolidated late Quaternary deposits and the underlying Tertiary volcanic and sedimentary rock (Fig. 46). Both the unconsolidated material and the surface on which it was deposited reveal a vertical displacement with the south side up. Moreover, a scarp with a vertical component of 1.2 m associated with the 1957 earthquake faces north in this area (Fig. 33). Hence, we suppose that the vertical offset in 1957, rather than being a freak occurrence, reflects a pattern of longer-term slip.

Figure 46. Photograph looking east along 1957 Gobi Altay earthquake rupture a few kilometers east of Ulaan Bulag (Fig. 31). Notice the large vertical separation at the surface forming the skyline (between downward-pointing black arrows) and the larger vertical offset of the Tertiary rock that forms the basement for young sediment deposited on it (between upward pointing open arrows). The basement is offset vertically about 12 m. With 1.2 m associated with the 1957 earthquake, this offset suggests that 10 earthquakes have occurred since the basement was planed flat and sediment accumulated on it. Notice also the vertical displacement in the foreground (oppositely pointed black arrows). A. C. stands next to the left arrow. (Photo by P. Molnar, July 1990.)

Two of us (A. C. and H. P.) recognized five similar sequences in the unconsolidated sediment on the north side of the scarp. In each, coarse material grades upward into finer material. These sequences may mark the occurrences of five earthquakes, each associated with a vertical component of slip that led to rapid downcutting in the hanging wall to the south and deposition of that material on the footwall to the north. The total thickness of this unconsolidated material is about 7 m. If material were deposited since 5 ka, 10 ka, or 20 ka, then recurrence intervals would be less than about 1,000 yr, 2,000 yr, or 4,000 yr. These numbers probably overestimate the lengths of recurrence intervals, because these deposits cannot record earthquakes that occurred after the stream incised and exposed this material. If the average horizontal component of slip for these events were 6 m as in 1957, then the average slip rate would be at least 6 mm/a, 3 mm/a, or 1.5 mm/a, respectively.

The vertical separation of the base of the unconsolidated material is 12 m (Fig. 46). If this difference were built by repetitions of earthquakes similar to that in 1957, with vertical components of 1.2 m (Fig. 33), then 10 such earthquakes would have occurred. If deposition began at 5 ka, 10 ka, or 20 ka, then rates of recurrence would be 500 yr, 1,000 yr, or 2,000 yr. Again assuming the same ratio of strike-slip to vertical components (5:1) as for the earthquake in 1957, this would imply horizontal slip of 60 m. For intervals of 5 ka, 10 ka, or 20 ka, these yield average slip rates of 12 mm/a, 6 mm/a, or 3 mm/a.

The range of assumed dates for the deposition of unconsolidated material obviously does not permit accurate assessments of recurrence intervals or slip rates. Nevertheless, because of the flat surface on which the unconsolidated material was deposited, an important change in climate is probably required for the change from erosion to deposition. Such a change seems most likely since the last glacial maximum at about 20 ka. Hence, we deduce that recurrence intervals are of the order of 1,000 yr, and horizontal slip rates are between 1 and 10 mm/a.

The other area where one can see cumulative displacement that might also be late Pleistocene or Holocene in age is on the south flank of Ih Bogd, along the Gurvan Bulag rupture (Figs. 31 and 45). This rupture cuts an active and relatively smooth fan along most of its length. At its eastern end, however, dissected remnants of an older fan stand roughly 45 m above the surface of the present fan and on the hanging wall of the fault that ruptured in 1957 (Fig. 45). As above, if we assume that this fan is a remnant dating from 5 ka, 10 ka, or 20 ka, its height suggests a rate of vertical displacement of 9 mm/a, 4.5 mm/a, or 2.3 mm/a. If it was built by repetitions of earthquakes with vertical components of 3 m, comparable with that in 1957, then 15 such earthquakes occurred, with average recurrence intervals of 330 yr, 700 yr, or 1,300 yr, respectively. Obviously, there is no evidence requiring that this zone ruptured every time that the Bogd fault ruptured. Nevertheless, it is logical to assume that major changes in geomorphic processes are related to the same climatic changes.

Collectively, these crude calculations suggest a recurrence interval of the order of a thousand years and an average slip rate of a few millimeters per year.

Digression on the origin of Ih Bogd. Ih Bogd ("Great Saint," or "Great Elevated One") is an extraordinary mountain. Its summit is nearly flat over tens of square kilometers and is flanked by very steep sides (Fig. 47). Presumably this flat surface represents an elevated remnant of an older, lower erosional surface. Ih Bogd lies south of the segment of the Bogd fault that trends east-southeast and where vertical components of slip for the rupture in 1957 were greatest. Florensov and Solonenko (1963) reported an average ratio of vertical to strike slip of 1:3. Thus, for this east-southeasterly segment, it could have reached 1:2. Indeed, in the segments to the east and west, where the local strike is oriented more nearly east-west, the vertical components were smaller fractions of the total slip. Thus, Ih Bogd seems to define a block caught in a left-lateral shear zone and uplifted where a right step in the zone requires horizontal compression (Fig. 31). The same can be said of Baga Bogd to the east. Moreover, recall that the southwest side of Ih Bogd simultaneously underwent uplift relative to the surrounding area by slip on reverse faults dipping northeast beneath the range. Thus, instead of being tilted by thrusting faulting occurring on only one side, Ih Bogd seems to be rising without undergoing much tilt.

Ih Bogd does not seem to be isostatically compensated. A. P. Bulmasov (Chapter VI of Florensov and Solonenko, 1963) reported that Bouguer gravity anomalies over the entire Gurvan Bogd range are more positive than those in the lower areas to the north or south. No data were published. Such gravity anomalies suggest a mass *excess* beneath the range instead of the deficit that would exist if isostatic compensation of any kind existed. Therefore, the weight of Ih Bogd seems to be supported by two relatively strong blocks that slide past one another along the Bogd fault.

The rates of slip inferred above suggest that Ih Bogd is very young. The flat summit stands about 2,500 m above the flat areas to the north and south. Insofar as the mountain is totally uncompensated, the vertical components of slip on the flanks should equal amounts of uplift of the mountain with respect to the surroundings. If a Holocene rate of 4.5 mm/a for that slip described active uplift, Ih Bogd would have formed only since 600 ka! Similarly, for an assumed horizontal slip rate of 6 mm/a on the Bogd fault, and for vertical components a third or half as much, the mountain would have risen since only 800 ka or 1,300 ka. Thus, Ih Bogd may have formed since continental glaciation began in the Northern Hemisphere at about 2.5 Ma (Shackleton and others, 1984). It might have formed since 1 Ma. The summit may have been too low to have undergone alpine glaciation except near the end of the Pleistocene epoch. In fact, Florensov and Solonenko (1963) reported finding no evidence for glacial cirques or for moraines on Ih Bogd. The lack of glaciation may have contributed to the preservation of the flat summit.

Some curiosities of the 1957 surface rupture. In many localities, the fault scarp that formed in 1957 faces uphill. Thus, the faulting that occurred in 1957 would seem to be different from the longer-term displacements that formed the present topography.

In some such localities, as near where the Toromhon Overthrust approaches the Bogd fault (Fig. 32), differences in resistance to weathering probably are responsible for this unusual relationship. More rapid erosion of easily eroded Mesozoic clastic sediment in the hanging wall than of resistant Paleozoic granite in the footwall apparently maintains relief of a sense opposite to that formed by the vertical components of slip.

In other areas, such as along the Bogd fault a few kilometers west of the area just mentioned (Fig. 36) or near the northern end of the Toromhon Overthrust (Fig. 42), differences in rates of erosion do not seem to be responsible for the scarps facing uphill. In the case of the Toromhon Overthrust, the cumulative displacement may be so small that the relief is simply not related to the faulting. In the case of the Bogd fault, however, there seems no alternative to the inference that the surface rupture does not

Figure 47. Composite photographs of the flat summit plateau of Ih Bogd and its steep sides. (Fig. 59 of Florensov and Solonenko [1963].)

define the precise orientation and sense of slip that characterized the faulting over the past tens or hundreds of thousands of years. Because the horizontal component of the slip vector seems to be roughly constant where the local strike of the surface fault varies, it is easy to ascribe the variations in vertical displacements at the surface simply to spatial variations in the strike at the surface. Similarly, differences in the sense of vertical components between the surface rupture and the regional topography might reflect variations in time of the position of the surface trace above a more constant position and orientation of the fault at depth. Accordingly, the orientation of the surface trace apparently would be only a crude reflection of the orientation of the deeper fault. If this observation applies to other surface faults in the world, it suggests that inferences about segmentation of faults based solely on surface ruptures should be viewed with caution. Faults at depth may not be segmented in the same way.

A second peculiarity is a marked variation in the preservation of surface ruptures. Florensov and Solonenko (1963) noted that where the surface rupture crossed dry stream valleys, scarps as high as 1 to 2 m in January 1958 (Fig. 48) were essentially gone in the fall of 1958. We observed in 1990 that in most places where the rupture had cut unconsolidated material, the scarp had degraded to slope at about 30°, the angle of repose (Figs. 33 and 35). Thus, if one were to exploit the diffusion equation and the slopes of older scarps in such material to date them (e.g., Hanks and others, 1984), the appropriate initial condition for the Gobi Altay region might be a slope of about 30°.

The Gobi Altay earthquake is also noteworthy for the extraordinarily sharp surface ruptures that it produced in some places (Figs. 32a, 34, 40, and 41). Some of these ruptures are cut into consolidated sedimentary rock, and presumably the strength of that material maintains the scarps. In addition, however, scarps that face uphill (Figs. 34, 40, 41, and 42) also tend to be well preserved. These scarps have not undergone denudation resulting from runoff that accumulated over a large area. Thus, it is likely that the diffusion equation is inapplicable to them.

Figure 48. Photograph of the scarp along the Bogd fault taken by I. B. on 4 January 1958. View is southwest at the scarp where it crosses the Toromhon Sayr, a wide seasonal stream valley. In the summer of 1958 the scarp across the dry valley was nearly totally eroded (see Fig. 67 of Florensov and Solonenko [1963]). In 1990, there was no trace of the scarp at all within the streambed. The photo in Figure 35, looking east, was taken from essentially the same place as was this photo and shows the east edge of the Toromhon Sayr.

Other ruptures in the Gobi Altay

Evidence of various types along the Gobi Altay range attests to oblique thrust and left-lateral slip on east-west planes subparallel to the rupture zone similar to that associated with the Gobi

Altay earthquake of 1957. Such faulting is very clear on satellite imagery (e.g., Tapponnier and Molnar, 1979), and studies of some faults on the ground corroborate such inferences.

West of the 1957 rupture zone, disruptions associated with an earthquake in the following year and with prehistoric earthquakes reveal a similar pattern. The Bayan Tsagaan earthquake of 1958 April 7 in southern Mongolia, 45.1°N, 98.7°E (Fig. 4), might be best treated as an aftershock of the Gobi Altay earthquake. Because Khil'ko and others (1985, p. 66) give some independent information, we too discuss it. V. P. Solonenko (Florensov and Solonenko, 1963) reported surface deformation along an east-west–trending zone 15 km long along the southern slope of the Bayan Tsagaan Uul (Fig. 31) and associated it with this earthquake. Later, in 1977, Khil'ko and others (1985) found only 7 km of such deformation. Tension cracks oriented 020 to 030° imply left-lateral slip. Although Khil'ko and others (1985) reported vertical displacement of as much as 1 m, they considered the strike-slip component to be the greater. They assigned a magnitude of 6.9 to this earthquake.

Khil'ko and others (1985, p. 30–31) mapped recent deformation along a second rupture zone west-northwest of the Bayan-Tsagaan rupture, near the village of Chandman', 45.33°N, 97.98°E (Figs. 4 and 49). Largely left-lateral slip seems to characterize the deformation associated with a zone trending 095° and 50 km long. An eroded south-facing scarp no higher than 1 m marks the eastern segment of the fault, where it crosses a flat basin. A north-facing scarp can be seen farther west near the village of Chandman'. The most impressive features, however, seem to be large elliptical pressure ridges trending roughly 320° in the western part of the zone. These pressure ridges are as high as 2.5 to 3 m, as long as 10 to 15 m, and as wide as 8 to 10 m in the village of Chandman'. Khil'ko and others (1985) inferred a large component of left-lateral slip, but they gave no estimate of the amount of slip, except the statement that "it must be significant."

Figure 49. Soviet satellite photo of the eastern part of the Gobi Altay. The Nuuryn Höndiy (Valley of Lakes) crosses the photo along 45.7°N, through Böön Tsagaan Nuur in the upper right and Beger Nuur just west of the image near 45°40′ latitude. Notice the many roughly east-west–trending linear breaks in slope, which we presume to be active faults. The nearly east-west–trending Chandman' rupture crosses the area shown in the middle of the photo, north of the hill in the center and between east-west–trending ridges to the west, near 45°20′N. The Myangayn rupture lies north of the Chandman' rupture and is marked by a sharp east-west break in the topography near 45°32′N.

They assigned an age of 500 to 1,000 yr to this deformation.

North-northwest of the Chandman' rupture, Khil'ko and others (1985, p. 27–30) mapped another zone of recent faulting trending 095 to 110° for a distance of 80 km, 4 to 5 km south of the village of Myangayn (45.55°N, 97.37°E) (Figs. 4 and 49). This zone lies south of the Beger Nuur and Böön Tsagaan Nuur basins but north of the western continuation of the Bogd fault, which ruptured in the 1957 Gobi Altay earthquake. A scarp marks the fault along much of its length, but the sense of vertical slip varies along the fault. Near the middle of the length of surface rupture, the maximum height reaches 3 to 4 m and faces north. Some of us suspect that this scarp may be associated with more than one earthquake. Characteristic heights are 1 to 1.5 m, with scarps facing either north or south. As Khil'ko and others (1985) stated, the dominant displacement probably was left-lateral strike-slip, but an amount cannot be inferred reliably. They assigned an age of 1,000 to 1,500 yr to the scarp on the basis of its morphology.

Although these two ruptures (Chandman' and Myangayn; Fig. 4) are the only two examples of Holocene faulting that have been studied in this area, the satellite imagery (Fig. 49) reveals several sharply defined, linear features in the topography that probably mark faults active during Quaternary time.

In 1988, two of us (A. B. and M. G. D.) studied a very clear scarp, the Hujirtyn rupture, along the southern side of the Han Jargalantyn Nuruu (Figs. 24 and 25). A linear north-facing scarp, 2.5 to 3 m high, trends east-west for 26 km west of 45.75°N, 95.50°E. The scarp is cut into bedrock and into overlying deposits where they are present. In bedrock the fault surface dips south at 60° and is associated with steep slickensides and fracturing of adjacent rock. Where the fault cuts overlying sediment, the surface of the scarp dips gently ($< 20°$). No strike-slip component was recognized, and A. B. and M. G. D. think that it is minor. This scarp is noteworthy for lying on the south flank of the range (Fig. 24) and dipping south. Clearly the range was not built by slip on this fault.

Southeast of the Gobi Altay rupture, at the northern foot of the Baruun Sayhan Uul, the Baruun Sayhan reverse fault, a strand of the Gurvan Sayhan fault system, dips ($\delta \cong 30$ to $50°$) south beneath the range (Fig. 50). In one place, Devonian siliceous rock has been thrust over late Cretaceous red sedimentary rock on a plane dipping south at 30 to 40° (Tikhonov, 1974, p. 206). Farther southeast, the Gobi Altay is sliced by numerous sharply defined faults that are clear on the satellite photographs (Fig. 50). These include straight, east-southeast–trending faults that apparently are strike-slip as well as more arcuate traces that probably mark thrust faults (see northeast part of area shown in Fig. 50). We are aware of no evidence for recent ruptures on any of these faults but cannot dismiss the possibility.

Khil'ko and others (1985, p. 47–49) and Natsag-Yüm and others (1971, p. 55–56) mapped three small surface ruptures at the eastern end of the Gobi Altay (Figs. 4 and 50), where the topographic expression of active faulting is not at all impressive on the satellite photographs. They associated these ruptures with two earthquakes in this century.

They assigned two short surface ruptures to the Ünegt earthquake, of 1903 February 1, in southern Mongolia (43.3°N, 104.5°E, M = 7.8 [Richter, 1958], M = 7.5 [Khil'ko and others, 1985]) (Fig. 4). The longer rupture trends east-southeast for about 20 km, but measurable displacements were not found on this trace. A second surface rupture extends 8 to 10 km northeast from the west end of the first trace. For about 3.5 to 4 km of this segment, a scarp 1.5 to 2.0 m high faces northwest (Khil'ko and others, 1985; Natsag-Yüm and others, 1971). They could not measure an amount, or even the sense, of a strike-slip component, but they suspected it to be present. The principal isoseismal (9 on a scale of 12)—elliptical in shape, aligned west-northwest, and about 70 km long—surrounds the two rupture zones.

It is difficult to assign meaningful values to the relevant source parameters of the earthquake. On the basis of regional strain associated with other earthquakes in Asia (e.g., Molnar and Deng, 1984; Tapponnier and Molnar, 1979), we assume primarily reverse faulting, on planes dipping 45 and 60° beneath nearby mountain ranges but also small strike-slip components consistent with northeast-southwest crustal shortening. From the intensity distribution, a length of 50 km seems appropriate for the main rupture. The magnitude suggests a seismic moment of about 5×10^{19} N m (Molnar and Deng, 1984), with an uncertainty of at least a factor of two. For the shorter, northeast-trending rupture, the average vertical component of 1.5 m, reported by Khilko and others (1985), and a length of 10 km yield a scalar moment of 1.5×10^{19} N m, with an uncertainty of at least a factor of two. These small moments make only minor contributions to the overall twentieth-century seismic strain, discussed below.

Khil'ko and others (1985, p. 67–69) also associated surface faulting along a nearby northwest-trending zone, 18 km in length, with the Buuryn Hyar earthquake, 1960 December 3, in southern Mongolia (43.11°N, 104.50°E, M = 7.0) (Figs. 2, 4, and 50). Vvedenskaya and Balakina (1960) reported a fault plane solution consisting of nearly pure thrust faulting with a northeasterly trending P-axis. Natsag-Yüm and others (1971, p. 62) observed primarily reverse faulting with the southwest block thrust over the northeast one by 0.15 m to 0.2 m. They also noted a small strike-slip component of 0.05 m to 0.1 m, shown as left-lateral on the map given by Khil'ko and others (1985, Fig. 3.19, p. 48). Oblique reverse and left-lateral slip of 0.3 m on a plane dipping southwest at 30° and striking 060° for 18 km corresponds to a scalar seismic moment of 7.1×10^{18} N m.

Summary. The entire east-west–trending zone across southwestern and southern Mongolia seems to be associated with oblique reverse and left-lateral faulting. Mountain ranges form block uplifts and slide horizontally past one another. Offset features apparently dating from late Pleistocene or Holocene time suggest an average slip rate near Ih Bogd of 1 to 10 mm/a and a recurrence interval of great earthquakes like that in 1957 of about a thousand years.

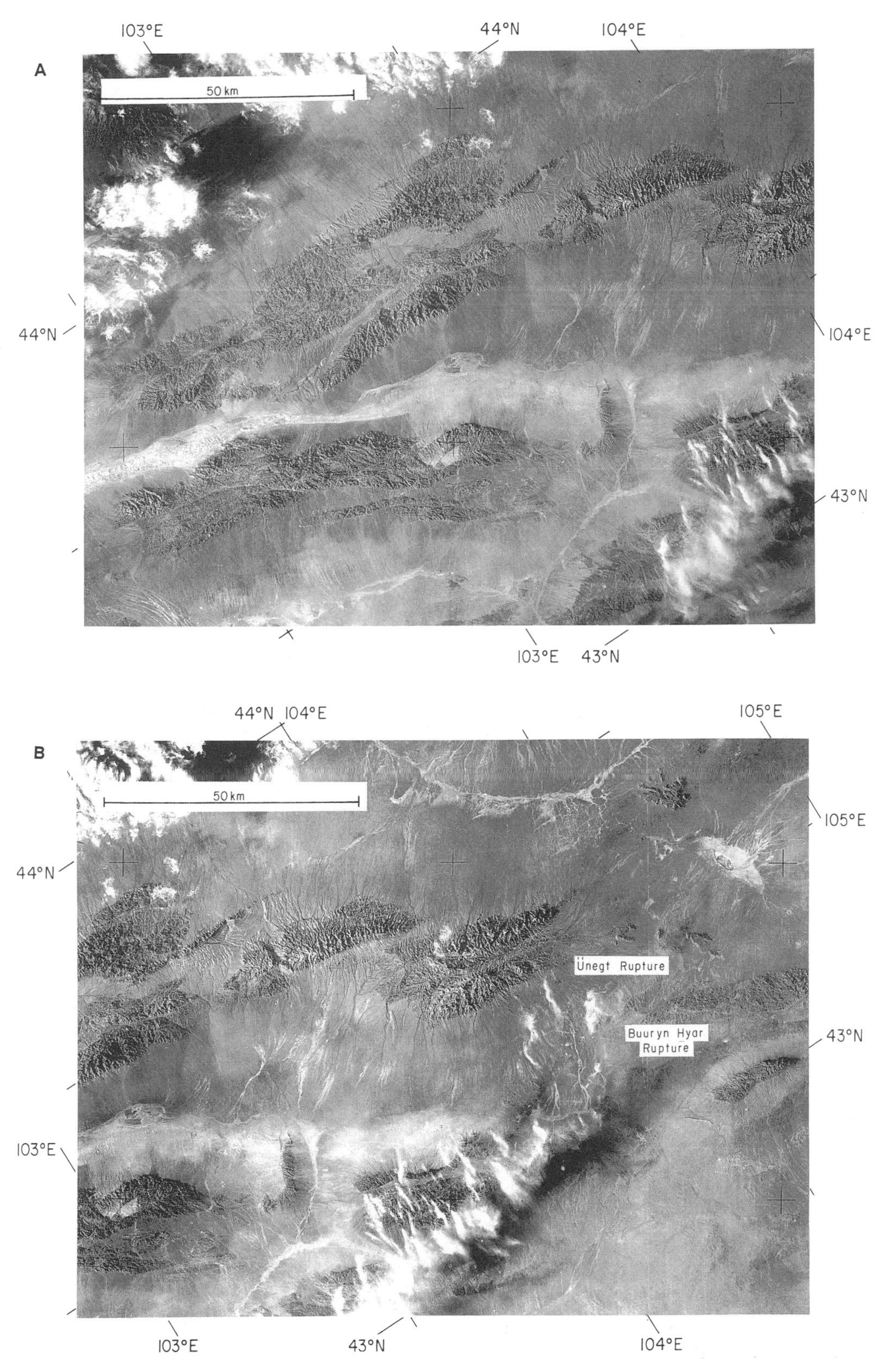

Figure 50. Overlapping Soviet satellite photos of the easternmost part of the Gobi Altay. Numerous sharp, easterly- to east-southeasterly-trending breaks in the topography, particularly in a, define what we presume are active fault traces. Ironically, the 1903 Ünegt and 1960 Buuryn Hyar earthquakes occurred in the eastern part of b, where surface deformation is perhaps the least impressive of any in the photos.

Because no evidence for prehistoric earthquakes with ruptures comparable to that of the 1957 Gobi Altay earthquake has been recognized, we cannot learn much from a sum of the products of average displacements times rupture lengths, as we did for the Mongolian Altay. That of the 1957 earthquake dominates the sum of 1.5×10^3 km · m. Nevertheless, if great earthquakes comparable to the 1957 Gobi Altay earthquake occurred somewhere along the entire 600-km length of the Gobi Altay at average intervals of 500 or 1,000 yr, then slip associated with such earthquakes would imply left-lateral strike-slip rates of 5 mm/a or 2.5 mm/a and a smaller north-south component of shortening (of roughly 2 or 1 mm/a). Such estimates ignore the smaller earthquakes, but obviously the greatest source of uncertainty is the guess of recurrence intervals. Clearly, some segments may rupture with great earthquakes and others only with small earthquakes. Rates of left-lateral slip of between 1 and 10 mm/a and shortening at roughly 2 mm/a seem reasonable.

THE HANGAYN NURUU AND CENTRAL MONGOLIA

The Hangayn Nuruu, or Hangay (Fig. 2), is a broad upland within the western part of central Mongolia. It is bounded and cut by active faults of different orientations and senses of movement. Maximum elevations in the Hangay exceed 3,000 m and decrease gradually toward the west, southwest, and south. Its boundaries with the Ih Nuuryn Hotgor to the west and the Nuuryn Höndiy to the south are not sharp (Figs. 2, 3, and 4). Devyatkin (1974, p. 185–186; 1975, 1981) reported that Oligocene red argillaceous sandstone and Oligocene and Neogene "plateau-basalts" cap parts of the southern and western periphery and hence define a well-developed late Cretaceous–Paleogene erosion surface. He used this surface as a "reference elevation bench mark" ("opornyi vysotnyi reper") over the Hangay to infer an uplift of at least 2,000 to 2,500 m of remnants of this surface relative to the surroundings.

The primary structure of the Hangay is a broad gentle dome, but within the highland, shallow basins separate secondary, or local, uplands (Devyatkin, 1975). In some areas the surface seems only gently warped and otherwise undeformed. In others, fault scarps bound basins. One of the better-studied depressions within the Hangay is the Bayan Hongor system of basins, 250 km long and 8 to 10 km wide. Electrical sounding revealed depths to the basement of 150 m in the Bayan Hongor basin itself and 240 m in the neighboring Bayan Tsagaan basin (Devyatkin, 1975, p. 276). The relatively thin sediment fill concurs with the basin's being a minor structure. The system is bounded by an important fault, the Bayan Hongor fault (Fig. 3), that is of Paleozoic age but reportedly rejuvenated in Plio-Pleistocene time. At one locality along the fault, where it crosses the Taatsiyn Gol (near 46.5°N, 101.2°E), Devyatkin (1974, p. 190) reported tightly folded early Cretaceous and Paleogene rock and Proterozoic metamorphic rock thrust southward over Oligocene sediment on a steeply (60 to 70°) northward-dipping reverse fault. (See also Tikhonov [1974, p. 200–201].) Elsewhere within the higher parts of the Hangay, however, scattered normal faulting seems to have occurred in Cenozoic time.

Normal faulting in the Hangay. Khil'ko and others (1985, p. 36–40) mapped a major recent surface rupture, the *Egiyn Davaa* scarp, 47.0°N, 99.6°E, across the Hangay where elevations exceed 2,000 m (Fig. 4). Three main segments collectively define a northeasterly trending zone of normal faulting. Throughout its 42-km length, the scarp faces northwest or west. The northeastern and southwestern segments trend northeast, but that in the middle trends more nearly north-south. The northeastern segment is a nearly continuous and straight rupture through Neogene basalt. At the base of one part of this segment, slickensides plunging about 35 to 40° toward 330 to 340° corroborate the predominantly dip-slip movement, but the orientations of tension cracks imply a small component of left-lateral slip. The southwestern segment is more discontinuous. In both segments, the height of the scarp ranges from 1 to 4.5 m, with a mean of 3 to 3.5 m in the northeast and 3.5 to 4 m in the southwest. Although the heights in the middle segment are lower, about 1.5 to 2 m, Khil'ko and others (1985) noted that the rupture splays into two or three subparallel traces so that the cumulative displacement in that segment might also be 4 to 4.5 m. They listed a mean vertical offset of 2.5 m, but a larger value seems required by their description.

From the fresh, steep walls of tension cracks (10 to 15°), Khil'ko and others (1985) assumed recent faulting and assigned an age of 300 to 500 yr, only a little older than the Sagsay and Ar Hötöl ruptures. Natsag-Yüm and others (1971, p. 54) noted that Mongolian legends told of a large earthquake in this general area during the life of Avday Sayn Haan, who lived in the second half of the sixteenth century, and suggested that this rupture should be associated with that earthquake. Khil'ko and others (1985, p. 76) offered a date of 1570.

Zorin and others (1982) noted that scattered normal faults cross the Hangay. On our brief traverse across the southwestern part of the Hangay, we traced one very prominent northwest-trending, northeast-dipping normal fault for more than 30 km near 47°N, 98°E (Fig. 51). This fault traverses a high area, much

Figure 51. Photograph of a normal fault in the southwestern part of the Hangay, near 47°N, 98°E. View is south across the northwest-striking fault. Note the clear triangular facets in the shadow of afternoon light. (Photo by P. Molnar, August 1990.)

of which lies above 2,500 m. With only a distant view to guide us, part of this fault appears to be broken by a fresh scarp, but to our knowledge it has not been studied.

Eastern Mongolia seems to be much less active than the western half (Devyatkin, 1974, p. 186, 194). This inference is corroborated by prominent active faulting seen on the Landsat imagery of western but not eastern Mongolia (Tapponnier and Molnar, 1979). The eastern limit of prominent active faulting in the Hangay might be the rupture zone of the Mogod earthquakes of 1967.

Mogod earthquakes of 1967 January 5 and 20 (48.22°N, 102.90°E, M = 7.8, and 48.10°N, 102.96°E, M = 6.7). Khil'ko and others (1985, p. 67–71) and Natsag-Yüm and others (1971, p. 74–82) associated these with a right-lateral fault zone that trends north-south and a southeast-trending zone of reverse faulting emanating from the south end of the main rupture (Fig. 52). The principal isoseismal is elongated in the north-south direction (Natsag-Yüm and others, 1971). The fault plane solution of the main shock (January 5) shows primarily strike-slip faulting (Fig. 2) (Huang and Chen, 1986; Tapponnier and Molnar, 1979). The fault plane solution of the largest aftershock (January 20, 1967), however, shows primarily thrust faulting on a plane striking southeast (Fig. 2). Moreover, Huang and Chen (1986) found that the P and SH waveforms from the mainshock required at least three subevents, with the fault plane solution of the third, southernmost subevent also showing largely reverse faulting on a plane striking southeast.

Natsag-Yüm and others (1971) visited the epicentral area on 6 January 1967, one day after the mainshock occurred. Their description was sketchy and focused more on large vertical displacements (Figs. 53, 54, and 55) than on the only sparsely impressive strike-slip component (Fig. 56). Later Khil'ko and others (1985) described a relatively simple pattern in which primarily right-lateral slip characterizes the displacement along a northerly trending rupture zone, 36 km in length (Fig. 52). At the northern end of the zone, displacements are not large, and unequivocal surface faulting is not very clear. In the southern two-thirds of the zone, however, mole tracks with long axes oriented 140° and tension cracks oriented 040° attest to right-lateral faulting. Farther south, the maximum horizontal displacement was described as "not less than 2.5 m" (Khil'ko and others, 1985, p. 71). Subsequently, in 1989, one of us (R. A. K.) measured a 3.2-m right-lateral offset of one gully (Fig. 56). The strike-slip displacement seems to increase from north to south along a segment some 36 km in length. The average horizontal displacement seems to be about 1.5 (± 0.5) m. Vertical displacements along this segment are in general smaller, but consistently with the east side up (Fig. 52).

Deformation is complicated and diffuse at the southern end of this rupture. Emanating from the southernmost 1 km of this zone are two reverse faults that trend southeast (140°) and dip northeast. The northern of these can be traced only for about 1 km. The southern, which continues for about 9 km, is the more important. Among other complexities between these two reverse faults are two northerly trending, west-facing scarps that seem to reflect normal faulting (Figs. 52 and 53).

Khil'ko and others (1985) reported vertical displacements along the main southeast-trending segment as growing from 1.5 to 2 m in the southeast to a maximum of 3 to 3.5 m (and possibly to 5 m [Natsag-Yüm and others, 1971]) near the junction with the north-south, strike-slip segment. The amplitude decreases again along the northwesternmost 1 km of the rupture, probably because some of the slip is transferred to the parallel surface rupture to the north (Fig. 52). Moreover, part of the great height of the scarp may be due to the degradation of it where it faces downhill, as we noted above in the discussion of the surface ruptures south of Ih Bogd and associated with the 1957 Gobi Altay earthquake (Fig. 44). Where the rupture cuts relatively flat topography, the maximum height of the scarp is about 2 m (Fig. 54a). Scattered northerly trending tension cracks and less common easterly trending mole tracks suggest a right-lateral component along this reverse fault, but we saw no unequivocal evidence of displaced gullies or other features (Fig. 54b). Thus, in our opinion, the strike-slip component must have been small, < 0.5 m.

The reverse faulting associated with the Mogod earthquakes is also peculiar in that the rupture slices along the sinuous crest of the Tüleet Uul, sometimes facing downhill (Figs. 53 and 54b) but sometimes uphill and blocking northeasterly drainage (Fig. 55). Thus, cumulative slip on this fault has not created the existing topography. It is tempting to assume that this fault has not been active for a long time. Moreover, the northern end of the north-south–trending strike-slip rupture passes through the middle of a northerly trending valley that seems to be bounded on its east side by faults with normal components. Thus, from the topography of the region, it is unlikely that we could have predicted the location or sense of motion on either of the segments that ruptured in 1967. It is tempting to infer (only tentatively) a very recent change in the style of deformation in this area.

The seismic moments obtained from long-period body waves, from long-period surface waves, and from the strike-slip surface deformation are internally consistent. The first two of three subevents of the mainshock recognized by Huang and Chen (1986) show strike-slip faulting, with seismic moments of 1.2 and 2.4×10^{19} N m. Ignoring the somewhat different orientations, the sum of their scalar moments (3.6×10^{19} Nm) yields a total that lies between estimates based on long-period surface waves, assuming strike-slip faulting: 3.2×10^{19} Nm (Okal, 1976) and 3.8×10^{19} N m (Chen and Molnar, 1977). An average slip of 1.5 (± 0.5) m along a segment 36 km in length and extending to a depth of 20 km also yields 3.6 (± 1.2) $\times 10^{19}$ Nm.

An average slip of 3.0 (± 1) m on the segment of reverse faulting dipping 42° to a depth of 20 km corresponds to an average vertical component of about 2 m. If it extends to a depth of 20 km, the scalar moment is 2.7 (± 0.9) $\times 10^{19}$ N m. This is more than twice Huang and Chen's (1986) estimate of the seismic moment for the third subevent of the mainshock: 1.2 $\times 10^{19}$ N m. The principal aftershock, for which reverse faulting

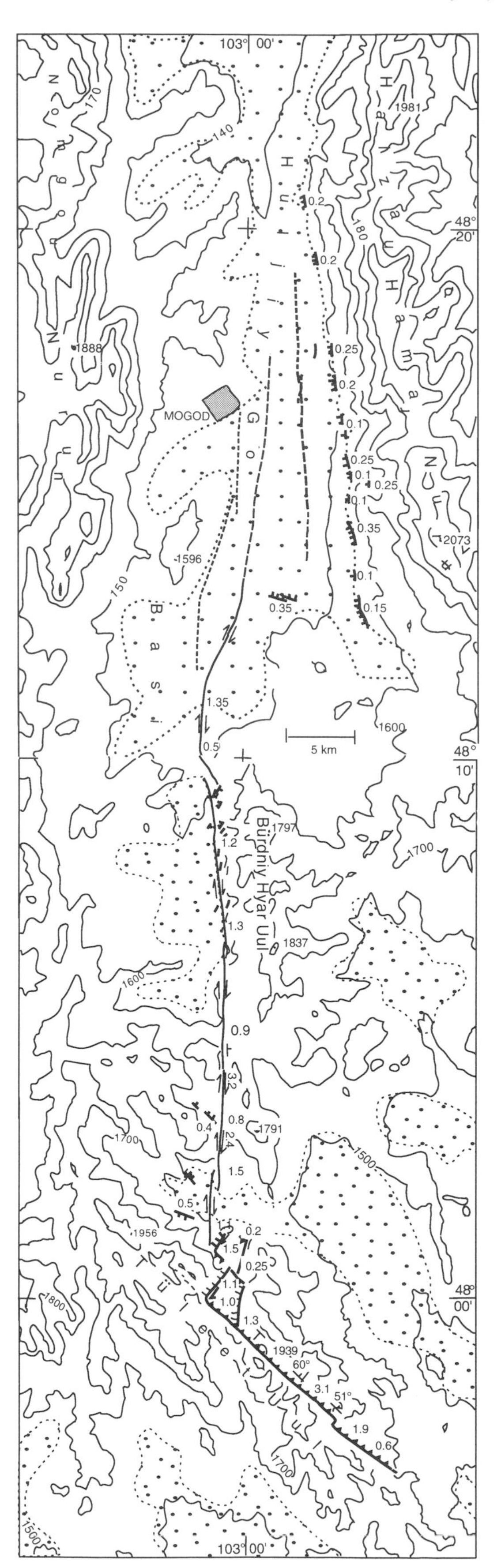

Figure 52. Map of the region surrounding Mogod earthquake rupture zone, based on unpublished work of R. A. Kurushin and V. V. Ruzhich in 1989. Fine lines show elevation contours in meters, and dark lines show surface ruptures. Teeth and hatching on the ruptures point in the downdip directions, with teeth showing reverse and hatching normal faulting. Numbers show heights of scarps, except where arrows give sense of strike-slip displacements. In two places northeasterly dips of 60° and 51° of the Tüleet Uul scarp along the Tüleet Uul, are also shown. Note that the main north-south–trending rupture passes through the center of the Hüljiyn Gol basin and not along its margins. The largest measured strike-slip offsets are from the southern part of this zone. At the southern end of the rupture, the Tüleet Uul rupture follows the crest of the Tüleet Uul.

Figure 53. Photograph of the thrust fault scarp (marked by the black arrows on the right) and a normal scarp (marked by the white arrow in the center) associated with the 1967 Mogod earthquakes. The curving normal fault scarp emanates from the more significant thrust scarp. View is toward the southeast along the southeast-trending Tüleet Uul rupture zone near its northwest end as shown in Figure 52. (Photo by P. Molnar, August 1990.)

also occurred and $M_0 = 0.5 \times 10^{19}$ N m (Huang and Chen, 1986), and other aftershocks might have contributed to the displacements observed along this segment. One of us (I. B.), however, was among the group who visited this surface rupture on 6 January 1967, two weeks before the main aftershock, and he has observed no change in the height of the scarp since that day. Thus, the aftershocks may have deepened the rupture without affecting the displacement at the surface. Alternatively, the rupture may have extended to a depth of only about 10 to 15 km, for which the seismic moment based on field observations would be only 1.3 to 2.0×10^{19} N m.

Figure 54. Photographs of the thrust fault scarp associated with the 1967 Mogod earthquakes. Views are toward the northeast across the southeast-trending rupture zone in Figure 52. a, In this area, near where the dip of the fault is estimated to be about 60° (Fig. 52), the local terrain is nearly flat, and the height of the scarp gives a reliable measure of the vertical component of slip, about 2 m. I. B., 1.8 m tall, provides the scale. b, Here, just tens of meters west of the area shown in Figure 53, a gully sloping southwest is offset. Clearly the strike-slip component is small. (Photos by P. Molnar, August 1990.)

THE BULNAY AND TSETSERLEG FAULT SYSTEMS

The Bulnay fault (Bolnai, Hanhöhiy Uul, or North Hangay fault) is a major strike-slip fault, clearly visible on the Landsat imagery (e.g., Okal, 1977; Tapponnier and Molnar, 1979). This fault marks the northern edge of the Hangayn Nuruu (Fig. 2).

Figure 55. Photograph of the thrust fault scarp associated with the 1967 Mogod earthquakes. View is toward the northwest along the southeast-trending rupture zone in Figure 52 (bottom). This photo was taken just a few tens of meters northwest of that shown in Figure 54a. Notice that the vertical component blocks drainage flowing northeast. (Photo by P. Molnar, August 1990.)

Figure 56. Photograph looking east along a gully offset 3.2 m horizontally, and a much smaller amount vertically, by the 1967 Mogod earthquake. R. A. K. (left) stands in the upstream thalweg, and M. G. D. (right) stands opposite the downstream thalweg. L. G. (center) stands behind on the east side of the rupture. This is the maximum horizontal displacement reported for this earthquake. (Photo by P. Molnar, August 1990.)

According to Zonenshain (1973, p. 223), there have been 50 km of post-Paleozoic left-lateral slip on it. This fault is probably best known for the very long rupture along it in 1905, associated with the second of two major earthquakes (Figs. 4 and 57). The first of them apparently ruptured the Tsetserleg fault (Figs. 3, 4, and 57), which trends northeast and lies north of the Bulnay fault. Beginning in mid-September 1905, Russian physical geographer A. V. Voznesenskii visited much of the area of surface faulting. We rely on his assignments of the different ruptures to the two earthquakes. He based these assignments on reports of local people, including both nomads and more sedentary Buddhist monks living in monasteries no longer extant (Voznesenskii, 1962).

Let us discuss the surface ruptures assigned to these earthquakes in their chronological order and then consider older ruptures along the zone.

Tsetserleg earthquake of 1905 July 9

This earthquake (49.5°N, 97.3°E, M = 8.4 [Richter, 1958], M = 7.8 [Khil'ko and others, 1985]) is associated with a zone of surface faulting trending approximately 060° for a distance of about 130 km. Surface displacements were variable in sense and amount, leading to a complicated pattern. Khil'ko and others (1985, p. 49–53) reported a left-lateral component of as much as 2.5 m along the eastern half of the rupture, which is marked by a continuous south-facing reverse scarp, 1.5 to 2.0 m high. They noted particularly common tension cracks in the central segment. At the southwest end, oblique reverse faulting occurred with up to 2.5 m of vertical displacement. The mapped fault trace is not straight. (The arrows on their map indicating right-lateral slip at the southwest end of the rupture seem to be misprints.)

We examined only a part of the southwestern end of the rupture, but our observations are qualitatively consistent with those of Khil'ko and others (1985). We measured a height of 0.85 m for a south-facing scarp (Fig. 58), where it locally strikes 060 to 070°. Casual observations suggested that where such a vertical component is present the local strike is more east-northeasterly or easterly than northeasterly. Correspondingly, where we measured a strike of 030 to 040°, no vertical component could be discerned (Fig. 59). We observed numerous examples of northerly trending tension gashes and easterly trending mole tracks that attest to left-lateral slip (Fig. 59), but we found no localities where we could measure an amount of strike slip.

From the structural relations, we presume that the horizontal component of the slip vector is oriented roughly 040°. Where the local strike is more east-northeasterly, the component of shortening across the fault manifested itself as oblique reverse faulting. Assuming such orientations, we can use the simple relationships shown in Figure 37 to make a rough estimate of the strike-slip component from the vertical component. For a northward dip of the fault of 63°, the component of convergence is half the vertical component. For a difference of 30° between the local strike and the azimuth of the slip vector, the strike-slip component should be twice the convergent component and hence equal

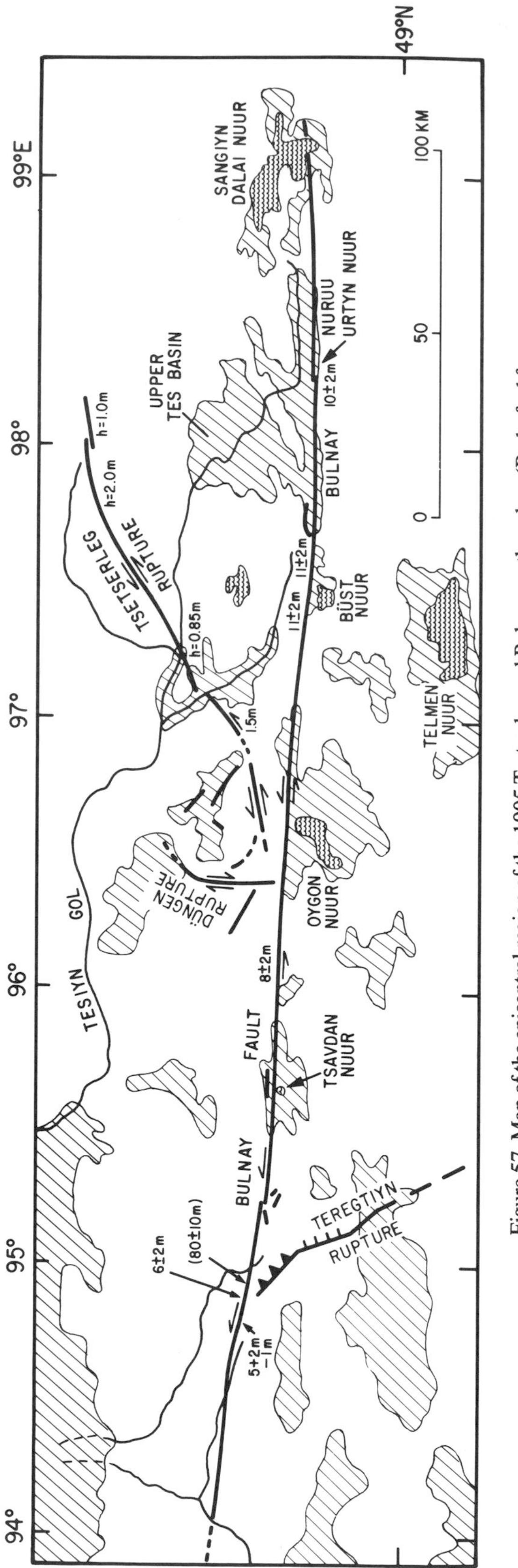

Figure 57. Map of the epicentral region of the 1905 Tsetserleg and Bulnay earthquakes. (Redrafted from Khil'ko and others [1985].) Locations of measured offsets and uncertainties (in meters) are denoted by the measured values. Strike-slip components are not designated. Vertical components are indicated by "h =." Diagonally hatched areas show sedimentary basins, and wavy texture shows lakes.

Figure 58. Photograph showing the vertical component of slip (0.85 m between black arrows) along the surface rupture of the 1905 Tsetserleg earthquake near the Tesiyn Gol (Fig. 57). View is north-northwest. R. A. K. provides the scale. (Photo by P. Molnar, August 1990.)

to the vertical component where the dip is 63°. Applied to the locality in Figure 58, these orientations imply a strike-slip component of 0.85 m. This calculation is too imprecise to place a tight constraint on the displacement, but it suggests that horizontal offsets as large as 5 m are unlikely.

Suppose that the average orientation of the horizontal component of the slip vector were 045° and that an average left-lateral strike-slip component of 2 (± 0.5) m occurred on a fault dipping north at 63° and trending 060° for 130 km. The average vertical component would be 1 m, the average slip on the fault would be 2.3 m, and the scalar seismic moment would be 2.2 (± 0.6) × 10^{20} N m. This estimate of the moment is considerably smaller than the range of 3 to 8 × 10^{21} N m deduced by Okal (1977) from long period Love waves recorded by Wiechert instruments at two stations. If Okal's moment is appropriate for the Tsetserleg earthquake, it suggests that the Tsetserleg fault was not the only one to rupture on July 9. Alternatively, the Wiechert instruments, whose natural periods commonly are only a few seconds, may not have been calibrated accurately enough at periods as long as 100 to 200 s to yield a reliable estimate of the seismic moment.

Bulnay earthquake of 1905 July 23

The magnitude, the average displacement and dimensions of the surface rupture, and the seismic moment of the Bulnay earthquake (49.2°N, 96.8°E, M = 8.7 [Richter, 1958], M = 8.2 [Khil'ko and others, 1985]) make it one of the largest known historic earthquakes in continental regions. Khil'ko and others (1985, p. 53–57) described three principal ruptures of very different orientations and noted that several other shorter segments (lengths < 10 km) also ruptured. We discuss only the three major faults here: the east-west–trending, left-lateral Bulnay fault, along which the principal rupture occurred; the north-south–trending, right-lateral Düngen rupture; and the southeast-trending Teregtiyn rupture, on which components of right-lateral and reverse faulting occurred.

Before discussing the displacements associated with this earthquake, it is worth calling attention to a few generalities. First, the rupture associated with this earthquake was large in all dimensions. In many areas the width of the disruption was tens of meters. In some cases displacements of a meter or more occurred on separate splays as much as a kilometer apart. Some tension gashes could be traced for more than 100 m. Others now comprise closed drainage basins with areas of many tens of square meters and with depths of meters (Figs. 60 and 61). Heights of

Figure 59. Overlapping photographs of large tension cracks and mole tracks associated with the 1905 Tsetserleg earthquake. View is northwest across the western end of the Tsetserleg rupture. Vehicles and people provide a scale. (Photos by P. Molnar, August, 1990.)

Figure 60. Photographs of large tension gashes and mole tracks associated with the 1905 Bulnay earthquake, just east of Büst Nuur. a, View west showing huge tension gash filled with water in left foreground. b, View east, from the same place, showing a large mole track in the foreground and alternating tension gashes and mole tracks farther east. (Photos by P. Molnar, August 1990.)

mole tracks commonly exceed 1 to 2 m, and their lengths also exceed tens of meters in some places. Moreover, because of the wide zone of surface faulting, tension cracks and mole tracks commonly overlap one another to make a mosaic of disrupted terrain several meters (or more) wide along the trace (Figs. 60 and 61). Although these observations consistently indicate left-lateral slip, they provide no evidence that allows quantitatively accurate estimates of offsets. Moreover, we found it difficult to match features offset along such segments of the fault.

The rupture is narrowest where it follows approximately the contours of topography with a moderate slope of a few degrees to about 30°. Steeper slopes are typically marked by slumping. Where the fault crosses flat terrain, the zone of disruption is widest and is nearly everywhere complex. Thus, most of the estimates of strike-slip displacements that we consider reliable were made where the scarp follows the contours of sloping topography. Virtually all were made from offsets of small streams and dry valleys oriented roughly perpendicular to the fault in such settings.

Some of us consider it likely that the widely distributed disruption in the valleys resulted from the fracturing of permafrost in these areas. In parts of northern Mongolia, permafrost is a few meters to more than 50 m thick (Gravis, 1974). We imagine that it could have been broken into blocks by the rupture of the subsurface in 1905. Freezing and melting along conduits formed in previous earthquakes may also been contributed to the complicated surface topography near the fault. Permafrost may be absent, however, where slopes are steep enough and especially where bedrock is near the surface. The narrow zones of rupture in such zones might, therefore, better reflect the dimensions of the fault at depth than the wide disruption in the flat basins.

If the complexities of the deformation were due to the fracturing of a layer of permafrost into blocks that moved rigidly with respect to one another, then "tension cracks" at the surface would be the result of separation between rigid blocks. Pressure ridges (and mole tracks) would mark localities where one rigid block was thrust over another. In such a case, the orientations of the tension cracks and pressure ridges would not necessarily bear a precise relation to the stress field at depth. As a result, the following discussion focuses more on measurements of displacements than on these complexities.

Bulnay fault. Most of the seismic moment can be associated with slip along an approximately 375-km-long segment of this east-west–trending, left-lateral strike-slip fault. Khil'ko and others (1985) and Trifonov (1985, 1988) reported measurements of displacements of the order of 3 to 6 m along different parts of the rupture. Their measurements differed from one another at most locations, with one reporting the largest offsets where the other reported rather small values. The averages of their measurements along the entire zone are both about 4 ± 1 m. Recognizing the difficulties in measuring offsets along this zone, we nevertheless conclude that their values generally underestimate the displacements by about two times.

The eastern end of the rupture lies just east of the lake Sangiyn Dalay and passes west beneath it and along the northern margin of the Bulnay Nuruu (see Voznesenskii and Dorogostaiskii's map, reproduced by Florensov and Solonenko [1963]). The trend is nearly due west, but there are small steps in the rupture. One left step of about 200 m near 98.3°E apparently manifests itself as a small pull-apart basin. A small lake, Urtyn Nuur, lies between the two strands.

Between Urtyn Nuur and where the Tsetserleg rupture approaches the Bulnay fault, the 1905 rupture is especially spectacular. About 2 km west of Urtyn Nuur, we measured a horizontal offset of 10 m, where the surface rupture displaced bedrock at the foot of the Bulnay Uul. At three other localities along the rupture farther west, to the area just west of the lake Büst Nuur, we measured 11-m displacements of small valleys

Figure 61. Photographs of large tension gashes and mole tracks associated with the 1905 Bulnay earthquake. Photos were taken (a) near Tsavdan Nuur (Fig. 57) and looking west and (b) near the junction of the Bulnay fault with the Teregtiyn fault and looking south-southwest. (Photos by P. Molnar, August 1990.)

(Fig. 62). In all cases, the measurements were uncertain by 2 m, and none was so clear that by itself it would constitute definitive evidence for such a large offset. Taken together, however, these four measurements suggest an average displacement of 10 to 11 m along this segment.

Along portions of this segment, the rupture is especially wide. Just west of where it crosses the Jarantay Gol, a splay to the north and then east follows the contours around the northern side of the Tesiyn Gol basin for several kilometers. The width of the rupture zone is at least 1 km there, and a large component (1 to 2 m) of normal slip can be seen on this splay. Moreover, the main rupture within the Tesiyn Gol basin is marked by a long, narrow graben, the Jarantay graben (Khil'ko and others, 1985), about 200 m in width, with large mole tracks and tension cracks along both sides. Difficulties in driving across the adjacent topography, with its numerous troughs that formed hexagons surrounding flat surfaces, made us acutely aware of deformation associated with freezing and thawing of the ground.

We suspect that an average displacement of about 10 m applies to the entire eastern half of the Bulnay rupture, extending roughly 200 km east to where the Tsetserleg fault approaches the Bulnay fault near the lake Oygon Nuur. Between the pass Möst Davaa,which lies roughly 15 km west of Büst Nuur, and the lake Oygon Nuur, the rupture crosses a relatively flat plain. There, disruption of the surface is as large as we saw, but we were unable to measure the offset.

In the segment of Bulnay fault between Oygon Nuur and where the Teregtiyn rupture approaches it, the displacement seems to have been slightly smaller. The fault follows the northern flank of a ridge for several kilometers, and several small gullies are clearly offset in a left-lateral sense. We measured one to be 8 ± 2 m (Fig. 63). Voznesenskii (1962, p. 45) reported an offset of 3 to 4 sazhens, approximately 30 km east of Tsavdan Nuur (Fig. 57) (a sazhen is an old Russian unit of distance equal to 2.1 m). Thus, his observation of 6.3 to 8.4 m is consistent with what we saw. We do not know the length of the rupture along which the average displacement is likely to be about 8 ± 2 m, but we suspect that it applies to the entire segment 100 km in length, between Oygon Nuur and where the Teregtiyn rupture approaches the Bulnay rupture.

Near that junction, the displacement seemed to us to be slightly less than that to the east. We measured two offsets, of 6 (± 2) m and 5 (+2/−1) m (Fig. 64), in the segment about 1 km long over which this junction is distributed. We also gained the qualitative impression that the disruption of the surface was greater to the east of this junction than to the west, but this may be a consequence of the narrow rupture passing across relatively steep terrain to the west. According to Khil'ko and others (1985), the rupture due to the earthquake in 1905 ends about 75 km west of this junction.

Figure 62. Photographs of 11-m offsets associated with the 1905 Bulnay earthquake. All three photos were taken near Büst Nuur (Fig. 57): a and b near one another to its west and c to its east. Views are to the south. a, The left-lateral offset of the gully is clear. b, A smaller gully than in a is offset. R. A. K. beneath left arrow stands at the head of the downstream thalweg, and A. C. beneath right arrow stands on the scarp above the base of the upstream thalweg. c, A wide stream is offset. Yu. Ya. V. (beneath right black arrow) stands at the edge of the right bank upstream of the fault, and I. B. (below left black arrow) stands on the same bank on the downstream side. (Photo a by H. Philip, and photos b and c by P. Molnar, August 1990.)

Figure 63. Photograph of a displaced gully along the Bulnay fault and associated with the 1905 Bulnay earthquake. The measured offset is 8 ± 2 m offset. R. A. K. stands in the fault zone and provides a scale. View is due south. The location is shown in Figure 57. (Photo using a 135-mm lens by P. Molnar, August, 1990.)

Trifonov (1988, p. 269) reported offsets of only 2.2 m and 3.8 m in the area near the junction with the Teregtiyn rupture. He assumed that larger offsets, not only here but also farther east where he reported only 5.5 m of slip and where we saw 11 m, reflect slip associated with more than one earthquake. We saw no indication of the smaller offsets that he measured. Perhaps his observations of very small offsets were made west of where the photographs in Figure 64 were taken.

Average displacements of 10 ± 2 m for 200 km, 8 ± 2 m for 100 km, and 5 ± 2 m for 75 km yield a seismic moment for the Bulnay rupture alone of 2.10 (± 0.49) × 10^{21} N m.

Teregtiyn rupture zone. V. A. Aprodov and O. Namnandorj discovered the northwestern part of this zone in 1958 (Aprodov, 1960). Later Khil'ko and others (1985) described in more detail ruptures extending for 80 km southeast from the main Bulnay fault, beginning approximately 75 km from its western end. Oblique reverse faulting characterizes the observed displacements on the northwestern segment, where Khil'ko and others (1985) measured vertical components that were typically 1.5 to 2.0 m and locally were 2.5 to 3.0 m. The northeast side moved up with respect to the southwest side. They reported right-lateral slip reaching 1.5 to 2.0 m and diminishing southeastward and a height of the scarp no more than about 0.5 m along the southeastern two-thirds of the rupture.

We visited only the northwestern quarter of the rupture, and a short section near the southeast end, but our observations are similar to those of Khil'ko and others (1985). At the northwest end of the Teregtiyn rupture, where it approaches the Bulnay rupture, relatively complex deformation is spread over a zone roughly 1 km wide. Deep tension cracks are oriented north-south (see Fig. 3.26 of Khil'ko and others [1985]), and several southwest-trending ruptures are present.

Figure 64. Photographs of offsets along the western segment of the Bulnay fault and associated with the 1905 Bulnay earthquake. Measured offsets are 6 ± 2 m in a (view south) and 5 (+2 or −1) m in b (view north). Locations are shown in Figure 57. (Photos by P. Molnar, August 1990.)

Beginning a few kilometers southeast of this junction, the rupture follows the southwest foot of a ridge of high terrain and is marked by a clear southwest-facing scarp trending 130° (Fig. 65). Where a stream has cut a cross section through the fault, the trace defines a clear northeast dip of about 50 to 70° (Fig. 66). Thus, the vertical component of about 1.3 m along this segment of rupture is due to reverse faulting. We were unable to measure a strike-slip component along this segment.

Farther southeast, the trace climbs to a pass; there the local strike is 135°, and no vertical component was observed. The scarp faces northeast where the trace crosses northerly sloping topography (Fig. 67). We assume that this scarp includes a right-lateral component of strike slip.

Faulting along this segment of the Teregtiyn rupture (Fig. 57) is not easily explained by slip of two blocks with a constant orientation of the slip vector, unless several meters of right-lateral slip occurred. The difference in the strikes of the segments with and without a reverse component of slip is only 5° to at most 10°. If the dip of the reverse segment were 63°, then, using Figure 37, 1.3 m of vertical slip would imply 0.65 m of convergent slip. Suppose that this component of convergence arose because of a bend in what is primarily a strike-slip fault. For variations in strikes of 5 or 10°, this convergent component would require 7.4 m or 3.7 m of strike slip. Such estimates obviously are very uncertain, because they are obtained by division of the vertical component by a small number, the tangent of the difference in the strikes of separate segments. Consequently, although the smaller estimate of the strike-slip component is certainly permitted by the scale of deformation that we saw, we are not confident either of the inference of several meters of right-lateral strike slip or of the basic assumption that the orientation of the slip vector is constant.

Between about 49.19°N, 95.09°E, and 49.12°N, 95.15°E, Khil'ko and others (1985) showed a trend of the fault of 150° instead of the more common 135° found to the northwest or 140° observed to the southeast. In this middle segment, the scarp consistently faces northeast, including an area where it crosses a flat basin. Thus, there must have been a vertical component of slip with a relative uplift of the southwest side, opposite to that observed to the northwest. This component of slip can be understood as a normal component if the slip vector is oriented 135 to

Figure 66. Photograph looking southeast along the Teregtiyn rupture of the 1905 Bulnay earthquake rupture, from the same place as in Figure 65. Oppositely pointing black arrows indicate the vertical component in the foreground. A diagonal black arrow points at the scarp on the hillside in the distance, where the trace fault shows the northeast dip. (Photo by P. Molnar, August 1990.)

Figure 65. Photograph showing the vertical component of offset (between arrows) on the Teregtiyn rupture near its northwestern junction with the Bulnay fault and associated with the 1905 Bulnay earthquake. View is toward the north. H. P. provides a scale. (Photo by P. Molnar, August 1990.)

Figure 67. Photograph, taken approximately 10 km southeast of areas in Figures 65 and 66, of the Teregtiyn rupture of the 1905 Bulnay earthquake. View is toward the southwest looking at the northeast-facing scarp. Note that uplifted side is opposite to that in Figures 65 and 66. (Photo by P. Molnar, August 1990.)

140°, corresponding to nearly purely right-lateral slip on the segments to the north and south with that orientation. In such a case, the height of 1 to 2 m of the scarp is also consistent with a few meters of right-lateral strike-slip displacement on the segments trending 135 to 140°.

In 1991, we crossed the southeast part of the rupture near 49°00′N, 95°15′E, where a low scarp about 1 m high faces west. The strike is 160°. We could see no clear evidence of strike-slip displacement, except for the straight trace. The variation in the sense of vertical offset along the entire Teregtiyn rupture, however, strongly suggests primarily strike-slip faulting. The scarp follows the west side of a series of very low northwest-trending hills, which are expressed on the topographic map by a series of closed contours (40-m interval). These probably are pressure ridges associated with strike-slip faulting.

For an average right-lateral component of 3 (± 1) m (trending 140°) along a segment 80 km in length and dipping vertically on average, the seismic moment would be 1.6 (± 0.5) × 10^{20} N m. This ignores the vertical components, but because of their opposite senses they would partially cancel in the estimation of the scalar moment.

Düngen rupture zone. Khil'ko and others (1985) described a third, short rupture zone that trends north for 20 to 22 km from the center of the Bulnay fault (Fig. 57). The principal manifestation of this rupture is a zone of very large tension cracks, as deep as 2 m, as wide as 3 to 5 m, and in some cases longer than 100 m (Fig. 68). Khil'ko and others (1985) inferred a substantial component of right-lateral slip, but they gave no quantitative estimate of the magnitude. We observed mole tracks in places, but we saw no evidence of a vertical component of slip on this zone.

From the dimensions of tension gashes, we think it likely that at least 1 m and possibly 2 m of right-lateral slip occurred. An average right-lateral slip of 1.5 ± 0.5 m on a vertical plane 22 km in length corresponds to a scalar seismic moment of 2.0 (± 0.7) × 10^{19} N m.

It is worth noting that Khil'ko and others (1985) show several other short traces of surface ruptures on their map. We, too, saw fresh scarps with vertical components of 1 to 2 m at the bases of several ridges for distances of a few kilometers. These additional ruptures seem to be more common near where the Tsetserleg, Düngen, and Teregtiyn ruptures approach the Bulnay rupture than elsewhere. They obviously suggest that the deformation was more complex than can be described with four roughly linear ruptures. Moreover, there could be other smaller ruptures that are yet to be discovered.

Older surface ruptures in the northern Hangay region

Khil'ko and others (1985) reported a short (8 to 10 km) oblique reverse fault strand near Malchin (49.7°N, 93.2°E), north and west of, but roughly parallel to, the fault that ruptured in the 1905 Bulnay earthquake (Fig. 4). They inferred left-lateral slip, roughly 300 to 500 yr ago, but they could not measure an amount of displacement. Devyatkin (1975, p. 271–272) described examples of late Cenozoic thrust faulting on roughly east-west–striking planes both north and south of the western termination of the Bulnay fault. On the Landsat imagery, a strand of reverse faulting can be recognized south of Uvs Nuur (Tap-

Figure 68. Photos of the Düngen rupture zone of the 1905 Bulnay earthquake (Fig. 57). a, View southwest along a tension crack more than 100 m long. H. P. stands to the left of the gash near its end. b, View south across a couple of tension gashes that define the Düngen rupture. A. C. stands adjacent to the distant gash, and H. P. stands in the nearer one. (Photo by P. Molnar, August 1990.)

ponnier and Molnar, 1979) (Figs. 2, 3, and 4), but farther west the north-northwest–trending right-lateral faults in the Mongolian Altay dominate the active tectonics.

Khil'ko and others (1985) also discovered two segments of the eastward continuation of the Bulnay fault (Fig. 4). A zone of surface faulting and disruption trending 080° for roughly 6 km begins near Bügseyn Gol (49.2°N, 99.6°E), about 25 km east of the rupture associated with the 1905 Bulnay earthquake. Northeasterly trending tension cracks attest to left-lateral strike-slip faulting. The western part of the disrupted zone is marked by a north-facing reverse fault scarp with a maximum height of 1.5 to 2 m. Khil'ko and others (1985) could not reliably estimate the strike-slip component, but they reckoned it to be "significant"—greater than the vertical component. From the eroded nature of the trace, they assigned an age of 1,500 to 2,000 yr to the surface faulting.

Two discontinuous segments, each roughly 20 km long and separated by 13 km, define a 53-km-long eastward continuation, starting near Züün Nuur (49.0°N, 99.8°E). At the western end a south-facing scarp 1.5 m high marks the rupture, but farther east the displacement seems to be purely horizontal. Tension cracks 40 to 50 m long, up to 5 m wide, and up to 2 m deep trend 050 to 055° (Khil'ko and others, 1985). The orientation of the eastern part varies from 080 to 085° in the west to 070 to 075°. In places, the scarp is high, up to 4 m, but with a sense that varies along the fault. Northeasterly tension cracks, 3 to 5 m wide and no more than 1.5 m deep, corroborate a left-lateral sense. Trees growing in the tension cracks for 250 yr require a greater age of faulting, and Khil'ko and others (1985) inferred an age of 300 to 500 yr.

Farther east, the Bulnay fault seems to become disrupted in grabens and discontinuous northeasterly trending segments with large components of normal faulting (Tapponnier and Molnar, 1979).

Average slip rates and rates of recurrence on the Bulnay fault

Near the intersection of the Teregtiyn and Bulnay ruptures, we saw one valley clearly offset tens of meters (Fig. 69). We measured a 74-m offset of the very flat center of the valley and an 89-m offset for the eastern edge of the valley. The upper reach of the valley, 100 to 200 m from the fault, is steep, with a sharp V-shaped cross section. Hence, we presume that it formed by postglacial erosion. If so, then this valley formed no earlier than the end of the last glaciation: 20 ka (Bard and others, 1990). If 80 m of slip occurred since 20 ka, 10 ka, or 5 ka, it would imply an average slip rate of 4 mm/a, 8 mm/a, or 16 mm/a. The corresponding average recurrence intervals for slip of 6 m at that locality would be 1,500 yr, 750 yr, or 375 yr. The absence of a record of major earthquakes in the last few hundred years casts doubt on the fastest slip rate and shortest recurrence interval, but an average slip rate of several millimeters per year and an average recurrence interval of the order of 1,000 yr seem plausible.

Figure 69. Photograph of the Bulnay fault near where the Teregtiyn fault approaches it. View is toward the north. The Bulnay fault enters the area in the photo on the left just behind the yurts. The valley shown in the center of the photo is displaced left-laterally about 80 m (between open arrows). Farther from the fault the valley has a V-shape, and hence is almost surely postglacial in age. (Photo by P. Molnar, with 135-mm lens, August, 1990.)

Trifonov (1985, 1988) reported that his estimates of offset stream gullies and small ridges occur in multiples of 5.5 m along a 15-km-long segment of the main east-west–trending segment of the fault. He concluded that slip has occurred in large earthquakes with such displacements. We did not visit the segment where he made these observations. If his measurements were reliable, they would imply that our estimate of a larger displacement in the adjacent segment (about 8 m) and Voznesenskii's (1962) of 2 to 3 sazhens are unreliable or unrepresentative of slip in 1905. Recognizing that small grabens or tension cracks formed along the rupture zone in association with largely strike-slip faulting, Trifonov (1985) assumed that the ages of organic material that fill such depressions can be used to date previous earthquakes. With such logic, he assigned ages to the last seven events before the one in 1905 and deduced a recurrence interval of 600 yr. We cannot share his confidence in such a precision in recurrence intervals without a more complete presentation of this data. Nevertheless, the suggestion of Holocene offsets that exceed those of 1905 by several times suggests that the slip rate is several millimeters per year.

THE HÖVSGÖL GRABENS AND THE BAIKAL RIFT SYSTEM

Three short grabens in northern Mongolia, the Hövsgöl graben system, named for the lake that fills the easternmost graben, trend roughly north from the eastern end of the Bulnay and Tsetserleg strike-slip faults. These three grabens comprise the southwestern end of the Baikal rift system. Lake Baikal, the deepest lake in the world, fills a segment of a rift system some 2,000 km in length, from northern Mongolia northeastward over much of southern Siberia. Despite local complexities, normal faulting and roughly northwest-southeast crustal extension characterize the deformation along the rift system.

Left-lateral slip on the Bulnay and Tsetserleg faults apparently is absorbed, at least in part, by east-west extension across these grabens. The Hövsgöl graben system, in turn, ends just west of the Tunka graben (Fig. 2), an east-west–trending zone of oblique normal and left-lateral slip at the southwest end of Lake Baikal (e.g., Sherman, 1978; Sherman and Ruzhich, 1973). The lake fills a northeasterly trending segment 600 km in length, but the morphology of the lake bottom reveals several en echelon deep basins separated by narrow ridges. Diverging from the lake, midway along it, is the nearly parallel, northeasterly trending Barguzin graben. The rift zone continues toward the east from the northeast end of the lake and north of the Barguzin graben through the Muya-Udokan region as a series of *en echelon* northeasterly trending grabens. These grabens are particularly clear on the Landsat imagery (Tapponnier and Molnar, 1979), but east of roughly 122°E the rift system becomes ill defined in both the topography and the seismicity and therefore difficult to trace.

Brief geologic history and deep structure of the Baikal rift system

The dramatic relief and virtually all of the faulting seem to have formed only in Pliocene and Quaternary time (since about 3 to 4 Ma), but the area encompassing the rift system apparently was tectonically active throughout much of the Cenozoic era (e.g., Florensov, 1968, 1969; Logatchev, 1968, 1974; Logatchev and Florensov, 1978; Logatchev and Zorin, 1987; Logatchev and others, 1983). Evidence for such tectonic activity includes (1) scattered Cenozoic basaltic rock over much of northern Mongolia (Devyatkin, 1974) and southern Siberia (e.g., Kisilev, 1987) and (2) a distribution of and variations in lacustrine and fluvial sediments that define depositional basins over much of the Baikal region (Logatchev, 1974; Logatchev and Florensov, 1978; Logatchev and others, 1983).

The predominance of Mio-Pliocene ages of the volcanic rock implies that the initiation of volcanism predates the creation of the present dramatic rifted topography (Kisilev, 1987). Although there are clear compositional variations in space and time in the Cenozoic volcanic rocks, they are dominantly alkali olivine basalt and hence derived from the mantle (e.g., Kisilev, 1987; Kisilev and others, 1978). As noted above, scattered distribution of the rocks over a zone some 1,200 km by 400 to 600 km (Kisilev, 1987) suggests that since 10 Ma or earlier, the upper mantle has been somewhat hotter than that beneath much of the rest of Asia (e.g., Zorin, 1981; Zorin and Osokina, 1984; Zorin and others, 1982). Moreover, most of the late Cenozoic volcanism has occurred outside of the rift (Kisilev and others, 1978). The mean elevation of 1,000 to 2,000 m of this region without large isostatic gravity anomalies and the absence of recent crustal shortening suggest that the high elevations are compensated by hot mantle, as for the Hangay region, and not by thick crust. Late P-wave arrivals of more than 1 s from distant earthquakes and for rays passing under the Baikal region (Logatchev and others, 1983; Zorin and Rogozhina, 1978) also suggest that the mantle lithosphere is thin (e.g., Logatchev and Zorin, 1987; Logatchev and others, 1983). The heat flow measured through the floor of the lake is locally very high (up to 140 mW/m^3) (Lysak, 1978) and therefore reveals local perturbations beneath the regions associated with the lithospheric extension. The distribution of high heat flow, however, does not correlate well with rifted topography, and heat flow in the region surrounding the rift valley is approximately normal (e.g., Lysak, 1978, 1984, 1987). Although this might suggest only a negligible thermal perturbation associated with the rift or with the regionally elevated terrain, Zorin and Osokina (1984) showed that because of the slow diffusion of heat from the base of the crust, heat flow measurements are not likely to detect such a recent thermal perturbation. Thus, both the present structure and the volcanic history suggest that the upper mantle of a broad region has been relatively hot since roughly 10 Ma.

There is a suggestion of Eocene continental sedimentation in basins aligned roughly parallel to the present rift, but the bulk of the Cenozoic sediment was deposited in Oligocene and more recent time and directly on Jurassic rock (Logatchev, 1968, 1974). Logatchev and his colleagues have reported thicknesses of Oligocene and Miocene sediment in such basins commonly reaching 1,000 to 2,000 m and locally 4,000 m in the deepest basin (Logatchev and Florensov, 1978; Logatchev and Zorin, 1987). The thicknesses of Oligocene and Miocene sediment seem to vary gradually instead of being localized in fault-bounded basins. Only the rare gravel and coarse sand suggest significant relief (Logatchev and Florensov, 1978). Thus one gains the impression that in the Miocene epoch the region was neither a stable lowland nor the unusually active region that it is today.

Tectonic activity accelerated dramatically in the Pliocene epoch; most of the present relief began to form at that time (Bazarov and others, 1974; Kurushin and others, 1966; Florensov, 1968, 1969; Logatchev, 1968, 1974; Logatchev and Florensov, 1978; Sherman and Ruzhich, 1973). Scattered outcrops of Oligocene and Miocene sedimentary rock on mountain tops and thick Plio-Quaternary sediment in deep basins attest to young tectonic displacements. The deep lake indicates that erosion and deposition cannot keep pace with the faulting, where vertical displacements exceed several kilometers.

The high level of seismicity, notably higher than that of the East African Rift or the Rheingraben, demonstrates continued activity (e.g., Golonetskii, 1977; Golonetskii and Misharina, 1978). Recent fault scarps are common along the rift system: in the Hövsgöl and neighboring grabens (Khil'ko and others, 1985), in the Tunka graben (Khromovskikh and others, 1975), along the margins of Lake Baikal (Solonenko and others, 1968), and in the Muya-Udokan region northeast of the lake (Khil'ko, 1966; Kurushin and Solonenko, 1966; Solonenko and others, 1966a, 1966b). Both the surface faulting and the fault plane solutions of large earthquakes attest to predominantly normal faulting, with T-axes and extension approximately perpendicular to the trend of the rift zone (Misharina, 1967, 1972; Tapponnier and Molnar, 1979; Vvedenskaya, 1961; Vvedenskaya and Balakina, 1960).

Thrust faulting, conjugate strike-slip faulting, and normal faulting perpendicular to the rift also occur locally in places along the rift system (Golonetskii and Misharina, 1978; Misharina and Solonenko, 1972; Misharina and others, 1983; Ruzhich and others, 1972), presumably in response to inhomogeneous deformation along the zone, but they are minor compared with the impressive normal faulting.

The Hövsgöl graben system in northern Mongolia

Three roughly parallel, northerly trending grabens mark a transition from the left-lateral Bulnay and Tsetserleg faults, in northern Mongolia, to the southwestern end of the Baikal rift system (Figs. 2 and 3). The lake, Hövsgöl Nuur, fills the easternmost of these grabens, but all three grabens are clear on the Landsat imagery (Tapponnier and Molnar, 1979). Devyatkin (1975, p. 278–282) reported that these grabens probably formed since Plio-Pleistocene time, with the central Darhad graben the earliest, the western Büsiyn Gol next, and finally the Hövsgöl graben.

Khil'ko and others (1985) described only two late Holocene surface ruptures from these grabens: one in the Büsiyn Gol graben (51.1°N, 98.0°E) and the other, the Jara Gol rupture, in the Darhad graben (51.4°N, 99.8°E) (Fig. 4). Both are characterized by normal faulting. The first was traced for a distance of 20 km, but no estimate of the displacement was given. Khil'ko and others (1985) traced a scarp 1.0 m to 2.5 m high along the second zone for about 5 km and in a direction 020°. They suggested that this short segment is part of a longer zone of discontinuous segments. They assigned ages of 500 to 1,000 years to reach.

Composite fault plane solutions of earthquakes at the northern end of the grabens indicate large strike-slip components, consistent with left-lateral slip on the easterly trending nodal planes (Misharina and others, 1983). Geologic estimates of strike-slip components of faulting are comparable with, or even exceed, normal components near the southwest end of Lake Baikal in the Tunka graben (Fig. 2). Both Holocene stream channels and basement structures have been offset left-laterally (Sherman and Ruzhich, 1973; Sherman and others, 1973). Sherman and Ruzhich (1973) reported 11 km of oblique normal and left-lateral slip in the Tunka graben. Hence, the Tunka graben seems to serve as a leaky transform fault between the Hövsgöl and Lake Baikal grabens (Sherman, 1978; Tapponnier and Molnar, 1979). In addition, the Tunka graben is one of the few places along the entire rift system where young volcanic activity is clear, if not very extensive.

Rates and amounts of opening of the Baikal rift system

Two simple arguments suggest that the rate of opening has been a few millimeters per year.

First, the rate of earthquake occurrence is consistent with such a rate. In this century, there has been one earthquake with a seismic moment of 10^{20} N m. The surface deformation (Solonenko and others, 1966a) associated with the Muya earthquake of 1957 June 27 (M = 7.9) implies that $M_0 = 1 \times 10^{20}$ N m (Molnar and Deng, 1984), which is consistent with the estimate of 1.4×10^{20} N m based on long-period surface waves (Chen and Molnar, 1977). Several other earthquakes with magnitudes only slightly smaller than that earthquake probably contributed a comparable amount to this century's total moment release. Moreover, because the Proval Bay earthquake of the last century (1862) was surely comparable to the Muya earthquake (Solonenko and others, 1968), it seems reasonable to assume that at least one earthquake of this size is typical for a century. Suppose, therefore, that the cumulative moment release per century is 3×10^{20} N m. Then, for a zone 1,500 km in length and a depth of faulting of 20 km, corresponding to fault widths of about 30 km, the rate of seismic slip is 2 mm/a. This yields a horizontal component, or an opening rate, of 1.4 mm/a. With a shallower depth of faulting or more earthquakes, this rate of opening could be 2 to 3 times greater. It is unlikely to exceed 5 mm/a, unless this century's seismicity is unrepresentatively low or most deformation occurs aseismically.

Second, the total amount of extension and its age offer a crude estimate of the rate of opening. Quantifying the total amount of displacement across the rift is difficult, but the likelihood of as much as several tens of kilometers seems remote. Largely from offsets on faults, Logatchev and others (1983) and Zorin (1981) suggested 15 km to a maximum of 25 km in the central Baikal graben, possibly decreasing to 10 to 15 km for the Tunka and Muya-Udokan grabens (Logatchev and Zorin, 1987). This accords with Sherman and Ruzhich's (1973) contention of 11 km for both the vertical and strike-slip components of oblique normal and left-lateral slip in the Tunka graben. Moreover, if Zonenshain's (1973) estimated 50 km of left-lateral slip on the east-west–trending Bulany fault occurred in Plio-Quaternary time, then 15 to 25 km of separation in a northwest-southeast orientation would not be surprising. Assuming a Pliocene initiation of the deep rift zone, since 4 to 5 Ma, the average rate of opening again would be millimeters per year. The apparent dying out of the rift system to the east suggests that the rate and amount of opening decrease northeastward (Zonenshain and Savostin, 1981).

SUMMARY AND DISCUSSION OF REGIONAL DEFORMATION

The dominant type of active faulting in western Mongolia and the adjacent area to the west in China is strike-slip: right-lateral on northerly to northwesterly trending planes and left-lateral on easterly or east-southeasterly trending faults. Large components of strike-slip faulting characterize essentially all of the major earthquakes that have occurred in this century. Earthquakes with $M \geqslant 8$ have occurred on four faults in three different parts of western Mongolia and its surroundings since 1900, and evidence of recent surface ruptures on other faults is abundant (Khil'ko and others, 1985). Evidence for both reverse (or thrust) and normal faulting also exists, and these styles of deformation might be more important than the seismicity of the twentieth century suggests. Nevertheless, slip on reverse or normal faults probably contributes only a fraction of the overall regional deformation. The strain field associated with this essentially conjugate strike-slip faulting can be described by roughly northeast-southwest shortening and northwest-southeast extension. Thus, it is logical to associate this shortening with convergence of India with respect to Eurasia (e.g., Molnar and Tapponnier, 1975). The extension appears to continue to the northeast, where it manifests itself as rifting and northwest-southeast extension across the Baikal rift system.

The average strain rate appears to be rapid and constitutes a substantial fraction of India's convergence with the rest of Eurasia. We can examine rates of deformation in two ways: from the seismicity of the twentieth century and from crude estimates of late Quaternary or Holocene slip rates on major faults. The difference between them provides a measure of the variability of seismicity over periods of 100 to 1,000 yr.

To use the seismic moments to estimate strain rates, we must estimate the seismic moment tensors for each major earthquakes in the region and then calculate their sum. The seismic moment tensor (Gilbert, 1970) is given by

$$M_{oij} = M_o (n_i b_j + n_j b_i)$$

where n and b are unit vectors normal to the fault plane and parallel to the slip vector. The directions given by i and j = 1, 2, and 3 are east, north, and up, respectively. The sum of the seismic moment tensors for earthquakes in this century is

$$M_{oij} = \begin{pmatrix} 0.23 & -1.99 & 0.18 \\ -1.99 & -0.65 & 0.07 \\ 0.18 & 0.07 & 0.42 \end{pmatrix} \times 10^{21} \text{ N m}$$

This tensor is symmetrical, and positive numbers in the top row correspond to east-west extension, right-lateral slip on an east-west plane (or left-lateral slip on a north-south plane), and uplift of the south side with respect to the north side on an east-west vertical plane, respectively. In the second row, positive numbers indicate right-lateral slip on an east-west plane (or left-lateral on its conjugate), north-south shortening, and uplift of the west side with respect to the east side on a north-south vertical plane. A positive number in the lower right indicates extension of the vertical dimension, or crustal thickening. We estimate the uncertainties in the quantities shown here to be roughly 30% of these values, ignoring ignorance of the depth range of faulting.

This sum of moment tensors can be converted into a strain rate by dividing it by twice the product of the shear modulus, the area of the region that includes the earthquakes, the thickness of the seismogenic layer, and the duration of time spanned by the seismic history (Kostrov, 1974). Consider a region 1.2×10^6 km^2 in area, corresponding to dimensions of roughly 1,200 km east-west and 1,000 km north-south. With an assumed thickness of 20 km and for a duration of 90 yr, the appropriate average strain rate tensor is

$$e_{ij} = \begin{pmatrix} 0.32 & -2.79 & 0.26 \\ -2.79 & -0.91 & 0.01 \\ 0.26 & 0.01 & 0.59 \end{pmatrix} \times 10^{-8}/\text{a}$$

The strain field is clearly dominated by left-lateral shear on east-west planes and right-lateral on north-south planes or equivalently by northeast-southwest shortening and northwest-southeast extension.

We may estimate the rate of northeast-southwest shortening across western Mongolia by multiplying this tensor by a vector representing the appropriate orientation and dimensions of the region: 1,100 ka east + 1,100 km north + 0 km up. This yields an average northeast-southwest convergence rate of 49 mm/a across western Mongolia in the twentieth century. This rate is indistinguishable from the rate of convergence between India and Eurasia of roughly 50 mm/a (DeMets and others, 1990). Even if the 30% uncertainty were associated with an overestimate of this twentieth-century strain rate, this shortening would be a very substantial fraction of the convergence rate between India and Eurasia. We doubt that this rate could apply for durations as long as millions, or even thousands, of years.

Similarly, the average twentieth-century rate of northwest-southeast extension is 40 mm/a. If the northwest-southeast extension in western Mongolia continued into the Baikal region, this rate would imply an opening of the Baikal Rift of 20 km in only 0.5 m.y., whereas geologic evidence suggests that rifting has been occurring for a few million years (Bazarov and others, 1974; Kurushin and others, 1966; Florensov, 1968, 1969; Logatchev, 1968, 1974; Logatchev and Florensov, 1978; Sherman and others, 1973).

There seems no escaping the inference that the seismicity of western Mongolia in this century has been unusually high. As discussed above, recurrence intervals for earthquakes like those in 1905 and 1957 are probably of the order of 1,000 yr. Similarly, occurrences of earthquakes like those in 1761 or 1931 in the Mongolian Altay probably occur less frequently than once per century. Thus, although major earthquakes will surely continue to occur in Mongolia in this century, the next great earthquakes like those in 1957 or 1905 may not occur for another century or more.

Estimates of late Quaternary or Holocene rates of slip on the major fault zones suggest crude regional strain rates corresponding to roughly 10 mm/a of northeast-southwest shortening. The record of earthquakes in the Mongolian Altay, not only for this century but also including events inferred from paleoseismic dislocations, suggests a rate of right-lateral slip, v, along the range of about 10 mm/a. Assigning an uncertainty is not easy, but a factor of two probably bounds the appropriate late Pleistocene–Holocene rate (5 mm/a $< v <$ 20 mm/a). Offsets of apparently late Pleistocene–Holocene landforms in the Gobi Altay region and along the Bulnay fault suggest rates of left-lateral slip of several millimeters per year on each. The sum of these rates also is about 10 mm/a, or between 5 mm/a and 20 mm/a. Thus, these rates of conjugate strike-slip displacements suggest northeast-southwest shortening and northwest-southeast extension at roughly 10 mm/a.

Accordingly, some 20% (± 10%) of India's penetration into Eurasia seems to be absorbed 2,000 to 3,000 km northeast of the Himalaya, the original plate boundary between the Indian and Eurasian plates when the boundary was narrow. Moreover, this approximate Holocene rate of extension, if extrapolated for a longer duration, is consistent with a Plio-Quaternary formation of the Baikal rift system.

Although the simple image of northeast-southwest shortening and northwest-southeast extension describes well the large-scale strain field, it obscures some important aspects of the active deformation in western Mongolia. The conjugate faulting is not symmetrical. Whereas the rates of left-lateral slip on the Bulnay fault and along the Gobi Altay may be comparable, slip along the Mongolian Altay is clearly much more rapid than the corresponding right-lateral shear in central Mongolia, reflected only by the Mogod earthquakes of 1967. Thus, the conjugate faulting cannot occur without some rotation of the interior of western Mongolia with respect to Siberia or to some other neighboring block.

In addition, the existence of conjugate faults that terminate against one another or that die out in thrust or normal fault systems requires that at least some of these fault systems rotate with respect to others about vertical axes. Slip on the Bulnay fault apparently is transformed into east-west extension across the Hövsgöl graben system. Left-lateral slip along the Tunka graben is transformed into extension in the Baikal rift. This deformation can be understood simply as slip between two blocks with an irregular boundary: Slip on faults with one orientation is transferred to those of another orientation. The Bulnay fault seems to die out to the west, however, before intersecting the northwest-trending Mongolian Altay. The right-lateral strike-slip faulting in the Mongolia Altay is associated with some crustal shortening at the northwest end of the range, but not so much that all of the strike-slip displacement is absorbed there. Ekström and England (1989) showed that terminations of strike-slip faults in a rigid block imply that such faults rotate with respect to the rigid block. Correspondingly, for right-lateral slip along the Mongolian Altay to be compatible both with its abrupt northwest termination in the relatively rigid Siberian block and with the eastward displacement of western Mongolia with respect to Siberia by left-lateral slip on the Bulnay fault, the right-lateral faults and the slivers of crust between them in the Mongolian Altay *must rotate counterclockwise* with respect to Siberia.

The right-lateral slip on the north-northwest–trending Mongolian Altay and the left-lateral slip on the east-southeast–trending Gobi Altay can also be understood as manifestations of counterclockwise rotation of the faults in the Mongolian Altay with respect those in the Gobi Altay. The areas northeast and southwest of these belts of active faulting may not yet have rotated much with respect to one another. Nevertheless, as deformation progressed and as the zone of active thrust and strike-slip faulting widened, the faults apparently rotated enough to make the presently very obtuse angle of about 135° between these two intersecting zones of opposite sense of shear. One might imagine that after continued deformation the Altay and Gobi Altay ranges will be parallel, and strike-slip faulting will cease or will be of the same sense along both ranges.

Thus, the simple image of the deformation of western Mongolia is that a roughly rectangular region adjacent to Siberia is evolving into a trapezoid. Left-lateral slip on the northern and southern margins maintains parallel boundaries while its western margin rotates counterclockwise. In a frame of reference attached to Eurasia, western Mongolia is undergoing regional left-lateral simple shear. The western margin (the Mongolian Altay) rotates counterclockwise and the eastern margin translates eastward, manifesting itself by opening of the rift systems to the north.

There are exceptions to the large-scale regional deformation described above. In particular, the interior of western Mongolia is not rigid; normal faulting characterizes the active deformation of much of the Hangay. Some of the resulting extension is oriented northwest-southeast, consistent with the regional strain field that would result from the conjugate strike-slip faulting surrounding the Hangay, but some normal faulting (Fig. 51) reflects northeast-southwest extension. This suggests that a uniform regional *stress* field does not characterize western Mongolia (e.g., Tapponnier and Molnar, 1979), even if a single large-scale regional *strain* field is consistent with most of the deformation. As is well known, variations in elevations, crustal thicknesses, and upper mantle structures perturb the stress field (e.g., Artyushkov, 1973). Gravitational potential energy is stored in elevated terrains, and this potential energy can be expended by crustal extension. Because more potential energy is stored in a region compensated by a hot upper mantle than one of the same elevation but underlain by a thick crust (e.g., Molnar and Lyon-Caen, 1988), normal faulting can occur in the Hangay while thrust faulting occurs beneath surrounding ranges of comparable height. Thus, the existence of a hot upper mantle beneath the Hangay alters what might otherwise be a relatively uniform regional stress field. The small amount of extensional strain, however, makes it only a negligible fraction of the regional strain field.

Given that the uppermost mantle beneath the Hangay seems to be a continuation of the hot upper mantle beneath the Baikal region (Zorin and others, 1982), the active strain field of western Mongolia seems to be a manifestation of two processes that affect the deformation of the crust. One obviously is the presence of such hot material beneath the Hangay, Hövsgöl, and Baikal regions. Because the regional northwest-southeast shortening is clearly more rapid than the extension across the Hangay, some process other than crustal extension in the Hangay must be responsible for this strain. The logical explanation is that convergence between India and Siberia induces the northeast-southwest compression of the intervening crust. Although the conjugate strike-slip faulting can be associated with this regional strain, the associated northeast-southwest shortening can explain neither the high elevations of the Hangay, Hövsgöl, and Baikal regions nor the normal faulting in the Hangay. Thus, we concur with Logatchev and Zorin (1987; Logatchev and others, 1983) that the collision of India with Eurasia *alone* cannot be responsible for rifting in the Bailal region or for the tectonics of western Mongolia. The emplacement of hot material, where normally there is cold lithosphere, must also have played a key role in the late Cenozoic tectonics of this area.

This interference of two apparently independent processes—upwelling from below and convergence between India and Siberia—is also suggested by the apparent youth of Cenozoic structures in western Mongolia. Extrapolations of the approximate Holocene rates of deformation are consistent with the Baikal rift being only a few million years old. The approximate Holocene rate for the Gobi Altay also implies that it formed in the last few million years. Ih Bogd, with its extremely flat summit, might have formed since 1 Ma. The ruptures associated with the 1967 Mogod earthquakes seem to bear little relation to the present relief, as if these ruptures follow very young faults. Thus, the abundant Pliocene and Quaternary conglomerate that surrounds elevated terrain in western Mongolia might, indeed, result from tectonic creation of relief, as is commonly inferred, and not from climate change, as may be the case elsewhere (Molnar and England, 1990). Unlike the Himalaya, Mongolia apparently has undergone intense tectonic activity for only a small fraction of the time since India collided with Eurasia in early Cenozoic time.

ACKNOWLEDGMENTS

Our fieldwork was supported by the Academies of Science of Mongolia and the U.S.S.R. through the Soviet-Mongolian Cooperative Geophysical Expedition. We thank P. C. England, M. C. McKenna, and R. E. Wallace for helpful reviews of the manuscript. The preparation of this manuscript was supported in part by the National Aeronautics and Space Administration under grant NAG5-795.

REFERENCES CITED

Aki, K., 1966, Generation and propagation of G waves from the Niigata earthquake of June 16, 1964. 2: Estimation of the earthquake moment, released energy, and stress-strain drop from G wave spectrum: *Bulletin of the Earthquake Research Institute of Tokyo,* v. 44, p. 73–88.

Aprodov, V. A., 1960, Seismotectonic observations in the region of the north Hangay earthquake of 1905 (MPR) (in Russian), *in* Questions of seismotectonics of pre-Baikal and adjacent territories: Bulletin of Soviet Seismology, Akademi Nauk USSR, no. 10, p. 90–97.

Artyushkov, E. V., 1973, Stresses in the lithosphere caused by crustal thickness inhomogeneities: Journal of Geophysical Research, v. 78, p. 7675–7708.

Avouac, J. P., Tapponnier, P., Bai, M., You, H., and Wang, G. 1993, Active thrusting and folding along the northeastern Tien Shan, and late Cenozoic rotation of the Tarim relative to Dzungaria and Kazakhstan: Journal of Geophysical Research (in press).

Baljinnyam, I., Mönhöö, D., Tsembal, B., Dugarmaa, T., Ad'yaa, M., and Bayar, G., 1975, Seismicity of Mongolia (in Mongolian): Ulaanbaatar, Academy of Sciences of Mongolia, 106 p.

Bard, E., Hamelin, B., Fairbanks, R. G., and Zindler, A., 1990, Calibration of the ^{14}C timescale over the past 30,000 years using mass spectrometric U-Th ages from the Barbados corals: Nature, v. 345, p. 405–410.

Bazarov, D. B., Antoshchenko-Olenov, I. V., Endrikhinskii, A. S., Ufimtsev, G. F., and Sizikov, A. I., 1974, Zabaikalia upland (in Russian), *in* Uplands of the Pribaikalia and Zabaikalia: Moscow, Nauka, p. 163–296.

Brune, J. N., 1968, Seismic moment, seismicity, and rate of slip along major fault zones: Journal of Geophysical Research, v. 73, p. 777–784.

Burrett, C. F., 1974, Plate tectonics and the fusion of Asia: Earth and Planetary Science Letters, v. 21, p. 181–189.

Chen, W. -P., and P. Molnar, 1977, Seismic moments of major earthquakes and the average rate of slip in Central Asia: Journal of Geophysical Research, v. 82, p. 2945–2969.

Dantsig, L. G., Dergachev, A. A., and Ivashchenko, A. I., 1965, An experiment applying the method of point sounding in processing seismological data for the Altay-Sayan region (in Russian), *in* Methods of seismic exploration: Moscow, Nauka, p. 92–99.

Dashzeveg, D., 1970, New data on the stratigraphy of middle Oligocene deposits of the MPR (in Russian), *in* Geology of the Mesozoic and Cenozoic of western Mongolia: Moscow, Nauka, pp. 37–43.

DeMets, C., Gordon, R. G., Argus, D. F., and Stein, S., 1990, Current plate motions: Geophysical Journal International, v. 101, p. 425–478.

Dergunov, A. B., Luvsandanzan, B., and Pavlenko, V. S., 1980, The geology of western Mongolia (in Russian): Moscow, Nauka, 196 p.

Devyatkin, E. V., 1970, Geology of the Cenozoic of western Mongolia (in Russian), *in* Geology of the Mesozoic and Cenozoic of western Mongolia. Moscow, Nauka, p. 44–102.

—— , 1974, Structures and formational complexes of the Cenozoic activated stage (in Russian), *in* Tectonics of the Mongolian People's Republic: Moscow, Nauka, p. 182–195.

—— , 1975, Neotectonic structures of western Mongolia (in Russian), *in* Mesozoic and Cenozoic tectonics and magmatism of Mongolia: Moscow, Nauka, p. 264–282.

—— , 1981, The Cenozoic of Inner Asia (in Russian): Moscow, Nauka, 196 p.

—— , 1982a, Paleogene (in Russian), *in* Geomorphology of the Mongolian People's Republic: Moscow, Nauka, p. 223–230.

—— , 1982b, Neogene-Anthropogene (stage of neotectonic activization) (in Russian), *in* Geomorphology of the Mongolian People's Republic: Moscow, Nauka, p. 230–245.

Devyatkin, E. V., and Shuvalov, V. F., 1990, The continental Mesozoic and Cenozoic of Mongolia (stratigraphy, geochronology, paleogeography) (in Russian), *in* Evolution of geologic processes and metallogenesis of Mongolia: Moscow, Nauka, p. 165–177.

Ding Guo-yu, ed., 1989, Atlas of active faults in China: Xi'an, Seismological Press, Xi'an Cartographic Publishing House, 121 p.

Ekström, G., and England, P. C., 1989, Seismic strain rates in regions of distributed continental deformation: Journal of Geophysical Research, v. 94, p. 10231–10257.

England, P. C., and Houseman, G. A., 1989, Extension during continental convergence, with application to the Tibetan Plateau: Journal of Geophysical Research, v. 94, p. 17561–17,579.

Ewing, M., and Press, F., 1959, Determination of crustal structure from phase velocity of Rayleigh waves. Part III: The United States: Geological Society of America Bulletin, v. 70, p. 229–244.

Florensov, N. A., 1968, The Baikal rift zone and some problems of its study (in Russian), *in* Baikal Rift: Moscow, Nauka, p. 40–56.

—— , 1969, Rifts of the Baikal mountain regions: Tectonophysics, v. 8,

p. 443–456.
Florensov, N. A., and Solonenko, V. P., eds., 1963, The Gobi Altai earthquake (in Russian): Moscow, Akademie Nauk USSR (English translation, U.S. Department of Commerce, Washington, D.C., 424 p., 1965).
Gilbert, F., 1970, Excitation of normal modes of the earth by earthquake sources: Geophysical Journal of the Royal Astronomical Society, v. 22, p. 223–226.
Golonetskii, S. I., 1977, Seismicity of Pribaikalia: History of its study and several results (in Russian), *in* Seismicity and seismogeology of Eastern Siberia: Moscow, Nauka, p. 3–42.
Golonetskii, S. I., and Misharina, L. A., 1978, Seismicity and earthquake focal mechanisms in the Baikal rift zone: Tectonophysics, v. 45, p. 71–85.
Gravis, G. F., 1974, Geographical distribution and thickness of multi-year frozen rock [permafrost] (in Russian), *in* Geocryological conditions of the Mongolian People's Republic: Moscow, Nauka, p. 30–48.
Gu Gongxu, and eight others, eds., 1989, Catalogue of Chinese earthquakes (1831 B.C.–1969 A.D.): Beijing, Science Press, 872 p.
Hangin, J. G., 1987, Basic course in Mongolian: Indiana University Publications, Uralic and Altaic Series, v. 73, 208 p.
Hanks, T. C., and Andrews, D. J., 1989, Effect of far-field slope on morphologic dating of scarplike landforms: Journal of Geoophysical Research, v. 94, p. 565–573.
Hanks, T. C., Bucknam, R. C., Lajoie, K. R., and Wallace, R. E., 1984, Modification of wave-cut and faulting-controlled landforms: Journal of Geophysical Research, v. 89, p. 5771–5790.
Huang, J., and Chen, W.-P., 1986, Source mechanisms of the Mogod earthquake sequence of 1967 and the event of 1974 July 4 in Mongolia: Geophysical Journal of the Royal Astronomical Society, v. 84, p. 361–379.
Khil'ko, S. D., 1966, Seismogenic structures of the Kodar system of activated faults (in Russian), *in* Recent tectonics, volcanoes, and seismicity of the Stanavoy Upland: Moscow, Nauka, p. 171–186.
Khil'ko, S. D., and Kurushin, R. A., 1982, Mongolian Altay (in Russian), *in* Geomorphology of the Mongolian People's Republic: Moscow, Nauka, p. 40–54.
Khil'ko, S. D., Kurushin, R. A., Kochetkov, V. M., Balzhinnyam, I., and Monkoo, D., 1985, Strong earthquakes, paleoseismogeological and macroseismic data, *in* Earthquakes and the bases for seismic zoning of Mongolia, Transactions 41, The Joint Soviet-Mongolian Scientific Geological Research Expedition: Moscow, Nauka, p. 19–83.
Khromovskikh, V. S., Solonenko, V. P., Kurushin, R. A., and Zhilkin, V. I., 1975, Seismogenic structures (in Russian), *in* Seismotectonics and seismicity of the southestern part of the eastern Sayan: Moscow, Nauka, p. 59–77.
Khutorskoy, M. D., and Yarmolyuk, V. V., 1990, Heat flow, structure and evolution of the lithosphere of Mongolia (in Russian), *in* Evolution of geologic processes and metallogenesis of Mongolia: Moscow, Nauka, p. 222–236.
King, G.C.P., and Vita-Finzi, C., 1981, Active folding in the Algerian earthquake of 10 October 1980: Nature, v. 292, p. 22–26.
Kisilev, A. I., 1987, Volcanism of the Bailal rift zone: Tectonophysics, v. 143, p. 235–244.
Kisilev, A. I., Golovko, H. A., and Medvedev, M. E., 1978, Petrochemistry of Cenozoic basalts and associated rocks in the Baikal rift zone: Tectonophysics, v. 45, p. 49–59.
Korina, N. A., and Nilolaeva, T. V., 1982, Hangay upland (in Russian), *in* Geomorphology of the Mongolian People's Republic: Moscow, Nauka, p. 87–109.
Kostrov, V. V., 1974, Seismic moment, and energy of earthquakes, and the seismic flow of rock (in Russian): Izvestiya, Akademi Nauk, USSR, Physics of the Solid Earth, no. 1, p. 23–44.
Kozhevnikov, A. V., Savin, V. E., and Uflyand, A. K., 1970, The history of development of the Hangay upland in the Mesozoic and Cenozoic (in Russian), *in* Geology of the Mesozoic and Cenozoic of western Mongolia: Moscow, Nauka, p. 151–169.
Kozhevnikov, V. M., Erdenebileg, B., Balzhinnyam, I., and Ulemzh, I., 1990, Structure of the earth's crust and upper mantle beneath the Hangay uplift (MPR) from dispersion of phase velocities of Rayleigh waves (in Russian): Izvestiya, Akademi Nauk, USSR, Physics of the Solid Earth, no. 3, p. 12–20.
Kurushin, R. A., and Solonenko, V. P., 1966, Seismogenic structure of the southern Muya zone (in Russian), *in* Recent tectonics, volcanoes, and seismicity of the Stanavoy Upland: Moscow, Nauka, p. 205–211.
Kurushin, R. A., Pavlov, O. V., and Khilko, S. D., 1966, Main neotectonic structures and activated faults (in Russian), *in* Recent tectonics, volcanoes, and seismicity of the Stanavoy Upland: Moscow, Nauka, p. 71–102.
Li Chunyü, Wang Quan, Liu Xueya, and Tang Yaoqing, 1982, Explanatory notes to the tectonic map of Asia: Beijing, Chinese Academy of Geological Sciences, 49 p.
Logatchev, N. A., 1968, Sedimentary and volcanic formations of the Baikal rift zone (in Russian), *in* Baikal rift: Moscow, Nauka, p. 72–101.
——, 1974, Sayan-Baikal-Stanavoy upland (in Russian), *in* Uplands of the Pribaikalia and Zabaikalia: Moscow, Nauka, p. 16–162.
Logatchev, N. A.,and Florensov, N. A., 1978, The Baikal system of rift valleys: Tectonophysics, v. 45, p. 1–13.
Logatchev, N. A., and Zorin, Yu. A., 1987, Evidence and causes of the two-stage development of the Baikal rift: Tectonophysics, v. 143, p. 225–234.
Logatchev, N. A., Zorin, Yu. A., and Rogozhina, V. A., 1983, Baikal rift: active or passive?—Comparison of the Baikal and Kenya rift zones: Tectonophysics, v. 94, p. 223–240.
Lysak, S. V., 1978, The Baikal rift heat flow: Tectonophysics, v. 45, p. 87–93.
——, 1984, Terrestrial heat flow in the south of east Siberia: Tectonophysics, v. 103, p. 205–215.
——, 1987, Terrestrial heat flow of continental rifts: Tectonophysics, v. 143, p. 31–41.
Misharina, L. A., 1967, Stresses in the earth's crust in rift zones (in Russian): Moscow, Nauka.
——, 1972, Stresses in the earthquake foci of the Mongolia-Baikal seismic zone (in Russian), *in* Fields of elastic stresses of the earth and focal mechanisms of earthquakes: Moscow, Nauka, p. 161–171.
Misharina, L. A., and Solonenko, N. V., 1972, On the stresses in the foci of weak earthquakes of Pribaikalia (in Russian): Izvestiya, Akademi Nauk, USSR, Physics of the Solid Earth, no. 4, p. 24–36.
Misharina, L. A., Mel'nikova, V. I., and Balzhinnyam, I., 1983, Southwest boundary of the Baikal rift zone from data on focal mechanisms of earthquakes (in Russian): Vulcanology and Seismology, no. 2, p. 74–83.
Mishen'kin, B. I., Mishen'kina, Z. R., and Seleznev, V. S., 1978, Structure of the earth's crust and upper mantle on the southwest flank of the Baikal rift (in Russian): Geology and Geophysics, no. 12, p. 3–13.
Molnar, P., and Deng Qidong, 1984, Faulting associated with large earthquakes and the average rate of deformation in central and eastern Asia: Journal of Geophysical Research, v. 89, p. 6203–6227.
Molnar, P., and England, P., 1990, Late Cenozoic uplift of mountain ranges and global climatic change: Chicken or egg? Nature, v. 346, p. 29–34.
Molnar, P., and Lyon-Caen, H., 1988, Some simple physical aspects of the support, structure, and evolution of mountain belts, *in* Processes in continental lithospheric deformation: Boulder, Colorado, Geological Society of America Special Paper 218, p. 179–207.
Molnar, P., and Tapponnier, P., 1975, Cenozoic tectonics of Asia: Effects of a continental collision: Science, v. 189, p. 419–426.
Nagibina, M. S., 1975, Some common regularities of development of Mesozoic and Cenozoic structures and Mongolian magmatism (in Russian), *in* Mesozoic and Cenozoic tectonics and magmatism of Mongolia: Moscow, Nauka, p. 290–296.
Natsag-Yüm, L., Balzhinnyam, I., and Mönhöö, D., 1971, Mongolian earthquakes (in Russian), *in* Seismic regionalization of Ulan Bator: Moscow, Nauka, p. 54–82.
Okal, E. A., 1976, A surface-wave investigation of the rupture mechanism of the Gobi Altai (December 4, 1957) earthquake: Physics of Earth and Planetary Interiors, v. 12, p. 319–328.
——, 1977, The July 9 and 23, 1905, Mongolian earthquakes: A surface wave investigation: Earth and Planetary Science Letters, v. 34, p. 326–331.

Philip, H., and Meghraoui, M., 1983, Structural analysis and interpretation of the surface deformations of the El Asnam earthquake of October 10, 1980: Tectonics, v. 2, p. 17–49.

Philip, H., Rogozhin, E., Cisternas, A., Bousquet, J. C., Borisov, B., and Karakhanian, A., 1992, The Armenian earthquake of 1988 December 7: Faulting and folding, neotectonics and palaeoseismicity: Geophysical Journal International, v. 110, p. 141–158.

Richter, C. F., 1958, Elementary seismology: San Francisco, W. H. Freeman, 768 p.

Rogozhina, B. A., Balzhinnyam, I., Kozhevnikov, V. M., and Vereshchakova, G. I., 1983, Peculiarities of travel times of P-waves from explosions in Nevada to seismic stations of Mongolia (in Russian): Geology and Geophysics, no. 4, p. 96–99.

Ruzhich, V. V., Sherman, S. I., and Tarasevich, S. I., 1972, New data on thrust faults in the southwest part of the Baikal rift zone (in Russian): Doklady Akademi Nauk USSR, v. 205, p. 920–923.

Sclater, J. G., Parsons, B., and Jaupart, C., 1981, Oceans and continents: Similarities and differences in the mechanisms of heat loss: Journal of Geophysical Research, v. 86, p. 11535–11552.

Sengör, A.M.C., Altiner, D., Cin, A., Ustaömer, T., and Hsü, K. J., 1988, Origin and assembly of the Tethyside orogenic collage at the expense of Gondwana Land, *in* Audley-Charles, M. G., and Hallam, A., eds., Gondwana and Tethys: London, Geological Society of London Special Publication 37, p. 119–181.

Shackleton, N. J., and 16 others, 1984, Oxygen isotope calibration of the onset of ice-rafting and history of glaciation in the North Atlantic region: Nature, v. 307, p. 620–623.

Sherman, S. I., 1978, Faults of the Baikal rift zone: Tectonophysics, v. 45, p. 31–39.

Sherman, S. I., and Ruzhich, V. V., 1973, Folds and faults of the basement: West Pribaikalia, Khamar-Daban and North Mongolia (in Russian), *in* Tectonics and volcanism of the southwest part of the Baikal rift zone: Moscow, Nauka, p. 24–35.

Sherman, S. I., Shmotov, A. P., and Medvedev, M. E., 1973, Rift structures (in Russian), *in* Tectonics and volcanism of the southwest part of the Baikal rift zone: Moscow, Nauka, p. 36–70.

Shi Jian-bang, Feng Xian-yue, Ge Shu-mo, Yang Zhang, Bo Mei-xiang, Hu Jun, and others, 1984, The Fuyun earthquake fault zone in Xinjiang, China, *in* Continental seismicity and earthquake prediction: Beijing, Seismology Press, p. 325–346.

Solonenko, V. P., Treskov, A. A., and Florensov, N. A., 1960, The catastrophic Gobi-Altay earthquake of 4 December 1957 (in Russian): Moscow, Government Scientific and Technical Publishing House, 48 p.

Solonenko, V. P., Kurushin, R. A., and Khil'ko, S. D., 1966a, Strong earthquakes (in Russian), *in* Recent tectonics, volcanoes, and seismicity of the Stanavoy Upland: Moscow, Nauka, p. 145–171.

Solonenko, V. P., Kurushin, R. A., and Pavlov, O. V., 1966b, Seismogenic structures of the Udokan system of activated faults (in Russian), *in* Recent tectonics, volcanoes, and seismicity of the Stanavoy Upland: Moscow, Nauka, p. 187–205.

Solonenko, V. P., and six others, 1968, Epicentral areas of early (preseismostatistical) earthquakes (in Russian), *in* Seismotectonics and seismicity of the rift system of Pribaikalia: Moscow, Nauka, p. 7–59.

Stein, R. S., and King, G.C.P., 1984, Seismic potential revealed by surface folding: 1983 Coalinga, California, earthquake: Science, v. 224, p. 869–872.

Tapponnier, P., and Molnar, P., 1979, Active faulting and Cenozoic tectonics of the Tien Shan, Mongolia and Baykal regions: Journal of Geophysical Research, v. 84, p. 3425–3459.

Tikhonov, V. I., 1974, Faults (in Russian), *in* Tectonics of the Mongolian People's Republic: Moscow, Nauka, p. 196–209.

Timofeev, D. A., and Nikolaeva, T. V., 1982, Gobi Altay and the Trans-Altay Gobi (in Russian), *in* Geomorphology of the Mongolian People's Republic: Moscow, Nauka, p. 65–87.

Trifonov, V. G., 1983, Late Quaternary tectogenesis (in Russian): Moscow, Nauka, 224 p.

—— , 1985, Peculiarities in the development of active faults (in Russian): Geotektonika, no. 2, p. 16–25.

—— , 1988, Mongolia—An intracontinental region of predominantly recent strike-slip displacement: Active faults (in Russian), *in* Neotectonics and contemporary geodynamics of mobile belts: Moscow, Nauka, p. 239–272.

Tsibul'chik, G. M., 1967, On the hodograph of seismic waves and the structure of the earth's crust in the Altay-Sayan region (in Russian), *in* Regional geophysical investigations in Siberia: Novosibirsk, Nauka, p. 159–169.

Vilkas, A., 1982, Sismicité et tectonique du Tien Shan: Ètude de quelques séismes par la méthode des sismogrammes synthétiques: Thèse de 3ème cycle, Universté de Paris XI, 172 p.

Voznesenskii, A. V., 1962, Investigation of the region of the Hangay earthquakes of 1905 in northern Mongolia (in Russian), *in* Materials from the Department of Physical Geographical Society of the USSR, Issue no. 1: Leningrad, 50 p.

Vvedenskaya, A. V., 1961, Peculiarities of the stress state in the foci of Pribaikal earthquakes (in Russian): Izvestiya, Akademi Nauk, USSR, Physics of the Solid Earth, no. 5, p. 666–669.

Vvedenskaya, A. V., and Balakina, L. M., 1960, Method and results of the determination of stresses acting in the foci of earthquakes of Pribaikal and Mongolia (in Russian): Bulletin of Soviet Seismology, Akademi Nauk USSR, no. 9, p. ;73–84.

Yang Zhang and Ge Shumo, 1980, Preliminary study of the fracture zone of the 1931 Fuyun earthquake and the features of neotectonic movements (in Chinese): Seismology and Geology, v. 2, p. 31–37.

Zaitsev, N. S., Zonenshain, L. P., and Mossakovskii, A. A., 1974, Some general aspects of Mongolian tectonics (in Russian), *in* Tectonics of the Mongolian People's Republic: Moscow, Nauka, p. 234–264.

Zhalkovskii, N. D., Tsibul'chik, G. M., and Tsibul'chik, I. D., 1965, Hodograph of seismic waves and thickness of the earth's crust of the Altay-Sayan folded region using data recorded from industrial explosions and local earthquakes (in Russian): Geology and Geophysics, no. 1, p. 173–179.

Zhang Pei-zhen, 1982, Surface ruptures associated with the 1931 Fu-yun, northeastern China, earthquake [Masters thesis] (in Chinese): Beijing, Chinese University of Science and Technology.

Zhang, Zh. M., Liou, J. G., and Coleman, R. G., 1984, An outline of the plate tectonics of China: Geological Society of America Bulletin, v. 95, p. 295–312.

Zonenshain, L. P., 1973, The evolution of central Asiatic geosynclines through sea-floor spreading: Tectonophysics, v. 19, p. 213–232.

Zonenshain, L. P., and Savostin, L. A., 1981, Geodynamics of the Baikal rift zone and plate tectonics of Asia: Tectonophysics, v. 76, p. 1–45.

Zorin, Yu. A., 1981, The Baikal rift: An example of the intrusion of asthenospheric material into the lithosphere as the cause of disruption of lithospheric plates: Tectonophysics, v. 73, p. 91–104.

Zorin, Yu. A., and Osokina, S. V., 1984, Model of the transient temperature field of the Baikal lithosphere: Tectonophysics, v. 103, p. 193–204.

Zorin, Yu. A., and Rogozhina, V. A., 1978, Mechanism of rifting and some features of the deep-seated structure of the Baikal rift zone: Tectonophysics, v. 45, p. 23–30.

Zorin, Yu. A., Novoselova, M. R., and Rogozhina, V. A., 1982, Deep structure of the Territory of MPR (in Russian): Novosibirsk, Nauka, 93 p.

Manuscript Accepted by the Society October 16, 1992

GLOSSARY OF MONGOLIAN WORDS USED IN GEOGRAPHIC NAMES

alt—gold
altan—golden
am—gorge, ravine
ar—behind, in back of
baga—small
baruun—west, right
bayan—rich
bogd—great, or elevated
bulag—spring
buur—bull camel (i.e., not castrated, like most, and very bad tempered)
buyant—virtuous
bürd—oasis
büs (büst)—belt
chih—ear
chihtey—having a quality of an ear (in this case, probably the shape)
dalay—sea, large lake; great
davaa—pass (in mountains)
dund—middle
gobi (properly, gov', for the i is only barely pronounced in Mongolian)—desert
gol—river
haan—(major) ruler (khan)
halzan—bald
hamar—low hills; nose
han—(minor) ruler
har (hara)—black
hoit—north
hot—town
hotgor—depression
hujir—salt
höl—light blue
höm—low area between the humps of a camel
höndiy (höndiyn)—valley
hötöl—a relatively low pass (lower than a davaa)
hövs—swim
hyar—plateau, elevated flat area
ih—great (often transliterated as ikhe or ikh)
jargal (jargalan)—happy
mörön—big river (only the Selenga qualifies in Mongolia)
mös (möst)—ice
nuruu—ridge, mountain range
nuur (nuuryn)—lake
ömön—south
övör—inner, bosom
sayhan—pretty, good
sayr—dry stream bed, arroyo
shar—yellow
sharga (shargyn)—dull yellow
shuvuut—bird
sonduul—a kind of long straight grass that grows in clumps
tajgar—short (applied to hair or grass)
tal—plain, steppe
tenger—heaven, gods
töhöm—saddle blanket
tsagaan—white
tsenher—blue
tulee—firewood
ulaan—red
urd—south, front
urt—long
us—water
uul—mountain range
üneg (ünegt)—fox
züün—east; left

Mongolian does not have prepositions, as do Indo-European languages; instead, such meanings are rendered by suffixes attached to the relevant nouns. The possessive case, as well as the preposition *of,* is formed by adding *-n, -yn,* or *-in* to words: Hangay*n* Nuruu is the Hangay range and Nuur*yn* Höndiy is the Valley of Lakes. The relationship denoted by the prepositions *to* or *in* is indicated by the suffix *-d* or *-t:* Mös*t* Davaa means icy pass (pass in ice) and Büs*t* Nuur is a lake in the form of a belt (there is a large island in its center). The idea of being with, having, or possessing is rendered by the suffix *-tay, -tey,* or *-toy:* The Chiht*ey*n Bulag is a spring, or group of springs, having some quality of an ear; in particular, the green, well watered area surrounding the springs is shaped like an ear.

NOTE ON PRONUNCIATION OF ROMANIZED MONGOLIAN SPELLINGS

Spellings of most Mongolian words used here are taken from the *Times Atlas of the World,* which employs a romanization somewhat different from that obtained from a transliteration of the Cyrillic spellings of Mongolian words. For other words we have tried to follow this romanization. For many years, Mongolian publications have used the Cyrillic alphabet, but the Mongolian government recently decided to return to the ancient Mongolian alphabet and script. Thus, it seems especially appropriate now to use a romanization that reflects as well as possible the sounds that are used by native speakers, instead of one transliteration—a romanization—of what is in effect another transliteration—Cyrillic—even if the Cyrillic transliteration reflects much more accurately the pronunciation than does the ancient Mongolian script (Hangin, 1987, p. vii).

Most vowel sounds are rendered acceptably by an English equivalent: *a* as in b*a*r, *e* as in l*e*t, and *i* as in *i*t. Mongolian, however, has three sounds that might be represented by the letter *o* in English but uses 3 different letters. The sound indicated by the letter *o* is like the *o* in h*o*t in British, but not American, English or like *ough* in *ough*t. The special letter *ö* is used to indicate a sound close to the *oo* in h*oo*f. The letter *u* is used to indicate a sound close to those of both *o* and *ö* and is made by shaping the mouth to say *u,* but trying to say *o* as in h*o*se; it resembles the English word *awe* but with a shorter duration than in English. Clearly the differences among *o, ö,* and *u* are subtle and not easily recognized by foreigners. To indicate the sound similar to that of *u* as in f*u*se, the letter *ü* is used. Mongolian also has a hard *i* sound, much like the hard *i* in Russian. This sound, indicated by *y,* is made by shaping the mouth as if to say *u,* but uttering an *ee* sound. The diphthongs *ay, ey, iy,* and *oy* are pronounced like *i* in b*i*te, *ei* in w*ei*ght, *ee* as in b*ee*t, and *oi* as in sp*oi*l. A repeated vowel in Mongolian lengthens the duration. Syllables with double vowels or diphthongs are accented: Ulaanbaatar is pronounced aw-**LAANg-BAA**t'r. In words with no diphthongs or no double vowels, there is no accented syllable, and Mongolian vowel sounds are very brief. Unaccented later vowels are nearly all pronounced the same, as in *uh* in *uh-huh.* Terminal vowels, as *-i* in Gob*i* (more correctly, gov'), are barely pronounced at all.

Most consonants are also rendered well by English equivalents. The letter *g* is sometimes pronounced like the hard *g* in English or like the much softer *g* as in Spanish (as in the word a*g*ua), hence almost like *w.* The sound indicated by the letter *h* is more guttural than the English *h,* but less so than the Russian *x* (kha), which is transliterated as *kh.* "Genghis Khan" should be written as "Chingis Haan," but in his case we have compromised, keeping the familiar "Khan." (Note also that a *haan,* like Chingis, is a much more important person than any ordinary *han.*) When an *h* forms the end of a syllable, the guttural kh-sound must be pronounced, as in *ih,* meaning great. A terminal *-n* is often pronounced like a very mild terminal *-ng* in English or somewhat like the nasal terminal *-n* in French, as in no*n.* "Ulaan" is pronounced as if it were written "ulaang." The Mongolian *r* is rolled somewhat, as in Russian or German, but not as much as in Italian and Spanish. The *v* in Mongolian is a compromise between *v* and *w.* When the letter *y* prcedes a vowel, it is pronounced as a consonant, as in *y*es. The letter *z* is closer to a *dz,* as in the Pinyin romanization of Chinese (e.g., Mao *Z*edong).

Printed in U.S.A.

Index

[Italic page numbers indicate major references]